New Approaches to Image Processing based Failure Analysis of Nano-Scale ULSI Devices

New Approaches to Image Processing based Failure Analysis of Nano-Scale ULSI Devices

Zeev Zalevsky

Pavel Livshits

Eran Gur

AMSTERDAM • BOSTON • HEIDELBERG • LONDON
NEW YORK • OXFORD • PARIS • SAN DIEGO
SAN FRANCISCO • SINGAPORE • SYDNEY • TOKYO
William Andrew is an imprint of Elsevier

William Andrew is an imprint of Elsevier
The Boulevard, Langford Lane, Kidlington, Oxford, OX5 1GB
225 Wyman Street, Waltham, MA 02451, USA

First published 2014

British Library Cataloguing-in-Publication Data
A catalogue record for this book is available from the British Library

Library of Congress Cataloging-in-Publication Data
A catalog record for this book is available from the Library of Congress

ISBN: 978-0-323-24143-4

For information on all William Andrew publications
visit our website at store.elsevier.com

This book has been manufactured using Print On Demand technology. Each copy is
produced to order and is limited to black ink. The online version of this book will show
color figures where appropriate.

Printed and bound by CPI Group (UK) Ltd, Croydon, CR0 4YY

Transferred to digital print 2013

CONTENTS

CONTENTS

PREFACE

The main aim of this book is to be a useful introductory and complementary technical textbook for engineers interested in extending their knowledge in the multidisciplinary field involving microelectronic circuitry fabrication and its failure analysis via imaging hardware and software.

The goal of the first part of the book (Chapters 1 and 2) is to acquaint researchers and engineers in the field of Advanced Metallization for Micro- and Nano-Electronics with novel image processing tools that might be relevant for improvement and analysis of blurred scanning electron microscope (SEM) and transmission electron microscope (TEM) images of metal ultrathin films' microstructures. We present an introduction to relevant image processing tools as well as a demonstration and a detailed explanation of a specific algorithm that actually obviates the need for time-consuming selection of SEM/TEM measurement conditions. The algorithm is different from all other existing methods, which are based solely on a priori knowledge about the inspected sample or the imaging system, in that it takes into account a priori knowledge about the microstructures' grain size and shape ranges. The capabilities of the algorithm are demonstrated on completely blurred high resolution (HR) SEM images of copper and silver films' microstructures, and on indistinct images of filled trenches/vias.

In the second part of the book (Chapter 3), we focus on superresolution approaches for microelectronics. Specifically, we present several novel numerical approaches for enhancing the resolution of low-resolution images that can be applied for failure analysis of microelectronic chips. The resolution improvement is based upon a numerical iterative comparison between a properly chosen image transform (such as the Radon transform) of a high-resolution layout image and the same transform of a low-resolution experimentally captured image of the same region of interest.

As one last comment, the authors would like to acknowledge and to thank Javier Garcia, Vicente Mico, Hamutal Duadi, Yoav Weizman, Philippe Perdu, Alex Inberg, Yossi Shacham-Diamand, and Arie Weiss for contributing to the material appearing in this book and for constructive technical interactions with the authors.

CHAPTER *1*

Introduction

1.1 BASICS OF IMAGE PROCESSING

In this chapter we focus on digital image processing. First, we are dealing in this book with optical image processing, which is discrete by
nature because the aperture of an optical setup determines the resolution (minimal distinguishable pixel size). Therefore, the image is not
really an analog, even if it originates from an analog function. Second,
most pre- and post-processing is done by a digital computer, giving
way again to digital image processing.

1.1.1 Introduction to Image Processing

Image processing (IP for short) is required in a variety of fields:
Clinical, Military, Astronomy, Microscopy, Security, Robot vision,
Agriculture, Meteorology, and more. IP is about improving images,
extracting essential information from images, and obtaining quantitative information from images, as well as assisting in diagnosis and displaying data in more productive ways. When we ask ourselves why we
need automatic image analysis, we can look at the pros and cons of
human experts versus computers. Humans are imprecise, indeterminis-
tic, expensive, and eventually get tired, causing the output to be

New Approaches to Image Processing based Failure Analysis of Nano-Scale ULSI Devices.
DOI: http://dx.doi.org/10.1016/B978-0-323-24143-4.00001-X

unreliable. However, they usually come with a large amount of knowledge and the capability of learning. Computers, on the other hand, have little knowledge and are not very clever, but they are precise, deterministic, cheap, and they never get tired.

Standard images are scalar images, meaning that each pixel in the image is represented by a single number, leading to a simple 2-D matrix representation. In complex images, each pixel is represented by a vector (displaying a direction), and so the matrix representation is built out of sub vectors, rather than single values. An example of a vector (complex) image could be an image describing a 4-D MRI blood flow [1].

One might ask himself what is an image? A formal definition [2] could be that an image is a two-dimensional rectilinear array of pixels (picture elements). In practice, we have a function mapping a 2-element vector (x- and y-coordinates) to a single-value gray level.

$$f(x, y): \mathbb{R}^2 \to \mathbb{R} \qquad (1.1)$$

The size of the matrix representing the image may vary from a low resolution image (e.g., $M \times N = 16 \times 16$ pixels) to a high resolution image (e.g., $M \times N = 2048 \times 2048$ pixels). Since the image does not contain continuous amplitude values, we may talk about depth resolution or the number of bits used (e.g., 4 bits allow 16 gray levels, while 8 bits allow 256 gray levels). Sampling is the term used to describe digitizing the coordinate values of our function (e.g., $f(x, y)$). Quantization is the term used for digitizing the amplitude values, $f(x, y)$. In practice, the sampling and quantization depend on the sensor arrangement that does the measurements.

Why is a sufficient sampling rate important? When we use too few pixels (resolution is too low), it is impossible to distinguish between similar images because the fine detail (higher spatial frequencies) is lost.

Why is a sufficient number of quantization levels important? If there's an insufficient number of intensity levels (small number of bits) we obtain false contouring, meaning that false contours or lines appear due to the insufficient number of intensity levels.

To overcome these problems, we need image interpolation. This means we use the data we know to estimate the data in unknown positions. The first option is the **nearest neighbor interpolation**, in which we use the known intensity level of the pixel closest to the one we are looking for as the intensity level of the pixel we are looking for. This is, of course, not very precise. The next option is **linear interpolation**. In this method, all the known values are connected by lines representing the original function by a set of linear subfunctions. This means that the value assigned to the required pixel is a weighted average of its two nearest neighbors. Since we are discussing 2-D image processing rather than 1-D signal processing, we might as well take advantage of the 2-D pixel arrangement. The simplest method here is a **bilinear interpolation**, taking into account the four nearest neighbors of the required pixel (2 on each axis). This is actually a 2-D extrapolation of the linear interpolation. There are other methods for interpolation of higher orders: in these methods more neighbors are involved and more complex curves are fitted to the known quantities.

We have to define the neighbors of a pixel. For the following 2-D case shown below (see Fig. 1.1), if P is the pixel located in the location (i, j) then the 4 nearest neighbors, also called the 4-neighborhood of P, are those marked in blue, while the 8 nearest neighbors, also called the 8-neighborhood of P, are those marked in blue and gray. We note the k-neighborhood of P by $N_k(P)$.

Next we define Connectivity and Adjacency (see Fig. 1.2). We say that r and s are **k-adjacent**, if their intensities are similar and $r \in N_k(s)$. Next we define a **k-path** as a sequence of points that are pairwise k-adjacent. Thus, a set of pixels $\{X\}$ are **k-connected** if for any two points in $\{X\}$ there is a k-path in $\{X\}$ between these two points. If so, then $\{X\}$ is a connected component.

When addressing connected components, it is important to separate the background components from the foreground components. In the following example, the number of connected components depends on the connectivity. On the left-hand side, the foreground is 4-connected

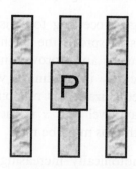

Figure 1.1 The neighborhood of a cell, introducing the 4 nearest neighbors (in blue) and the 8 nearest neighbors (in blue and gray).

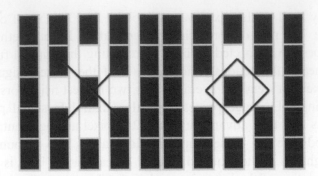

Figure 1.2 Connectivity and number of connected components: On the left-hand side, the foreground is 4-connected with 4 components and the background is 8-connected with one component. On the right hand side, the foreground is 8-connected with one component, and the background is 4-connected with 2 components.

and has 4 components, while the background is 8-connected and has 1 component. On the right-hand side, the foreground is 8-connected it has 1 component, while the background is 4-connected and has 2 components. It is also important to label connected components, since this allows us to automatically count objects in an image.

1.1.2 Histograms

Next we turn to histogram representation of data. We begin with the definition of a histogram. A histogram is a discrete function, $h(r_k) = N(r_k)$, where r_k is the k-th intensity value, and $N(r_k)$ is the number of pixels with intensity r_k. The histogram is normalized by dividing $N(r_k)$ by the number of pixels in the image ($N \times M$). The normalization turns the histogram into a probability distribution function, $P(r_k) = N(r_k)/(N \times M)$.

For example, below we can see histograms of four versions of the same image, from left to right: a light image, a dark image, a high-contrast image, a low-contrast image (see Fig. 1.3).

One of the basic image processing tools is histogram equalization (see Fig 1.4). The idea is to spread the intensity values to cover the whole gray scale; the result is an improved/increased contrast. Let us assume that r is the intensity in an image with L gray levels; thus $r \in [0, L - 1]$. Histogram equalization is mapping of the form $s = T(r)$, where r is the input gray level and s is the output gray level (the mapped value). Two conditions must be met:

I. $T(r)$ must be a monotonically increasing function in the region $[0, L - 1]$ and

II. If $0 \le r \le L - 1$ then $0 \le T(r) \le L - 1$.

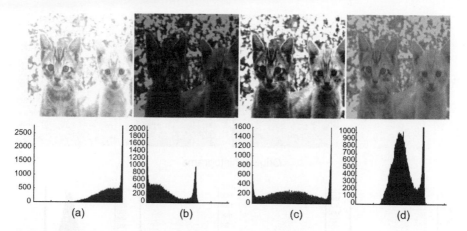

Figure 1.3 Images and histograms of (a) a bright image, (b) a dark image, (c) a high-contrast image and (d) a low-contrast image.

A good candidate for $T(r)$ is the **Cumulative Distribution Function** (CDF), which is the integral of the Probability Density Function (PDF). $CDF(r) = \int_0^r p_r(\eta)d\eta$. Since the PDF is nonnegative, the CDF is monotonically increasing, satisfying the first condition. Since $CDF(0) = 0$ and $CDF(L-1) = 1$, the output and input do NOT share the same range value, so we have to scale the result by $(L-1)$ to satisfy the 2nd condition.

$$s = T(r) = (L-1)CDF(r) = (L-1)\int_0^r p_r(\eta)d\eta \qquad (1.2)$$

Now we have to see how using such a mapping function affects the histogram, so we want to calculate the PDF of the output function. Using probability theory,

$$p_s(s) = p_r(r)\left|\frac{dr}{ds}\right|. \qquad (1.3)$$

As we already stated,

$$s = T(r) = (L-1)CDF(r) = (L-1)\int_0^r p_r(\eta)d\eta \qquad (1.4)$$

So

$$\frac{ds}{dr} = (L-1)\frac{d}{dr}\int_0^r p_r(\eta)d\eta = (L-1)p_r(r) \qquad (1.5)$$

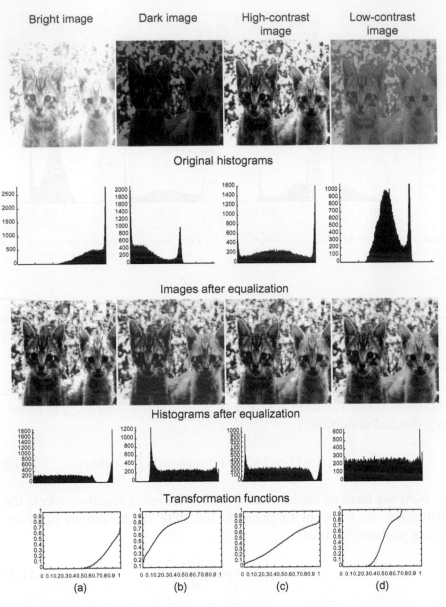

Figure 1.4 Images (1st row), histograms (2nd row), equalized images (3rd row), equalized histograms (4th row), and transformation functions (5th row) of (a) a bright image, (b) a dark image, (c) a high-contrast image and (d) a low-contrast image.

Meaning that

$$p_s(s) = p_r(r)\left|\frac{1}{(L-1)p_r(r)}\right| = \frac{1}{(L-1)} \tag{1.6}$$

where $0 \le s \le L - 1$, giving us a uniform distribution as required.

In the discrete case, the equalization is, of course, not perfect, but it spreads the intensity values to cover the whole gray scale and thus improves the contrast, in a routine that is very easy to implement. We recall that $N(r_k)$ is the number of pixels with intensity r_k, and write

$$s_k = \frac{(L-1)}{M \times N} \sum_{n=0}^{k} N(r_n) \tag{1.7}$$

The quality of equalization depends on the size of the matrix, as shown in the examples below, showing a MATLAB® simulation.

The two main features of the histogram are its mean value and its variance. The mean value is given by:

$$m = \langle r \rangle = \sum_{n=0}^{L-1} r_n \times p(r_n) \tag{1.8}$$

While the variance is given by:

$$\sigma^2 = \langle (r-m)^2 \rangle = \sum_{n=0}^{L-1} (r_n - m)^2 \times p(r_n) \tag{1.9}$$

The mean value is a measure of intensity or brightness and the variance is a measure of contrast.

1.1.3 Spatial Filtering

In spatial filtering, we take an input image $g_{in}(x, y)$ and use a filter to obtain an image $g_{out}(x, y)$. The filter contains a rectangular neighborhood, is mostly with odd dimensions, and has predefined operations. The filter generates a new value for the pixel in the middle of the neighborhood. For a 3 by 3 filter, the filter operation can be written in the form:

$$\begin{aligned} g_{out}(m,n) = & \omega(-1,-1)g_{in}(m-1,n-1) + \omega(-1,0)g_{in}(m-1,n) \\ & + \omega(-1,1)g_{in}(m-1,n+1) + \omega(0,-1)g_{in}(m,n-1) \\ & + \omega(0,0)g_{in}(m,n) + \omega(0,1)g_{in}(m,n+1) \\ & + \omega(1,-1)g_{in}(m+1,n-1) + \omega(1,0)g_{in}(m+1,n) \\ & + \omega(1,1)g_{in}(m+1,n+1) \end{aligned} \tag{1.10}$$

Or, in the more general case of a $(2A + 1) \times (2B + 1)$ filter:

$$g_{out}(m, n) = \sum_{i=-A}^{A} \sum_{j=-B}^{B} \omega(i, j) g_{in}(m + i, n + j) \qquad (1.11)$$

This means that the filter is actually shifted each time, and the locations overlapping the filter and the input image are multiplied, and finally all of these multiplications are summed up to yield the result. The output image is actually the **correlation** between the filter and the input image. If we compare correlation to convolution for the 2-D discrete case, and for real images, then correlation gives:

$$y(m, n) = \sum_{i=-A}^{A} \sum_{j=-B}^{B} x(i, j) h(m + i, n + j) \qquad (1.12)$$

While convolution gives:

$$y(m, n) = \sum_{i=-A}^{A} \sum_{j=-B}^{B} x(i, j) h(m - i, n - j) \qquad (1.13)$$

This is equivalent to first rotating the filter by 180 degrees and then performing a correlation. Note that correlation and convolution are identical if the filter is symmetric. The later equation actually demonstrates that for every linear shift-invariant system the output can be obtained by convolution between the input and the impulse response $h(m, n)$, where $h(m, n)$ is the output obtained when a discrete Kronecker delta is present at the input:

$$\delta(m, n) = \begin{cases} 1 & m = n = 0 \\ 0 & m, n \neq 0 \end{cases} \qquad (1.14)$$

We define the matrix representing the filter as the filter Kernel. The idea is to correlate or convolve an image with different filters in order to obtain different results (e.g., processed images).

Smoothing filters: average the intensities, thus resulting in a blurred image with less detail. For example:

$$\frac{1}{9} \begin{bmatrix} 1 & 1 & 1 \\ 1 & 1 & 1 \\ 1 & 1 & 1 \end{bmatrix}.$$

In this case, each pixel is replaced by the average of itself and its 8 closest neighbors. This is a normalized filter as the sum of the filter coefficient is 1 and the correlation is the same as the convolution due to symmetry. The bigger the filter size, the more blurred is the resulting image. The normalization is important since otherwise the intensities tend to get high and we might have a saturation effect.

Sharpening filters: these filters, on the other hand, enhance the regions of the image where intensities change fast, such as at edges. The basic 1-D derivation filter is given below:

$$[h_x] = [h_y]^T = \begin{bmatrix} 1 & 0 & -1 \end{bmatrix} \tag{1.15}$$

We can also use an arbitrary angle derivation, where given $[h_x]$ and $[h_y]$ the filter is

$$[h_\theta] = cos(\theta)[h_x] + sin(\theta)[h_y] \tag{1.16}$$

One common option is to use the **Prewitt Gradient Kernel** [3], where

$$[h_x] = \frac{1}{3} \begin{bmatrix} 1 & 0 & -1 \\ 1 & 0 & -1 \\ 1 & 0 & -1 \end{bmatrix} \text{ and } [h_y] = \frac{1}{3} \begin{bmatrix} 1 & 1 & 1 \\ 0 & 0 & 0 \\ -1 & -1 & -1 \end{bmatrix}; \text{ this means}$$

that we perform a derivative in one direction and smoothing in the other.

Another common option is the **Sobel Kernel** [4], where

$$[h_x] = \frac{1}{4} \begin{bmatrix} 1 & 0 & -1 \\ 2 & 0 & -2 \\ 1 & 0 & -1 \end{bmatrix} \text{ and } [h_y] = \frac{1}{4} \begin{bmatrix} 1 & 2 & 1 \\ 0 & 0 & 0 \\ -1 & -2 & -1 \end{bmatrix}.$$

Note that in the derivative filters shown above, the sum of all elements is zero: thus, if the image is constant, the result (derivative) is zero.

1.1.4 Fourier Analysis

Since processing in the Fourier plane has a major role in any image processing scenario, we will now briefly go over some of the most important properties related to the work of the French mathematician Jean Baptiste Joseph Fourier [5].

The starting point is that any periodic function can be expressed as a sum of sines and/or cosines, named the Fourier series. Even functions that are not periodic and have a finite area under a curve can be expressed as an integral of sines and cosines multiplied by a weighting function, named the Fourier transform. Both the Fourier series and the Fourier transform have an inverse operation.

Before we show how Fourier can make the life of the image processing expert easier, we now address the impulse function—perhaps the most important function for analyzing any system. The Unit impulse (Dirac delta function) is defined by:

$$\delta(t) = \begin{cases} \infty & t = 0 \\ 0 & t \neq 0 \end{cases} \tag{1.17}$$

With the constraint: $\int_{-\infty}^{\infty} \delta(t)dt = 1$.

Its main property is the shifting property: $\int_{-\infty}^{\infty} f(t)\delta(t - t_0)dt = f(t_0)$, with the special case for $t = 0$ of $\int_{-\infty}^{\infty} f(t)\delta(t)dt = f(0)$.

The Discrete unit impulse (for the 1-D case, the 2-D case was shown before) is defined as

$$\delta(m) = \begin{cases} 1 & m = 0 \\ 0 & m \neq 0 \end{cases} \tag{1.18}$$

With the constraint: $\sum_{m=-\infty}^{\infty} \delta(m) = 1$.

Here the shifting property means: $\sum_{m=-\infty}^{\infty} f(m)\delta(m - n) = f(n)$, with the special case for $n = 0$ of $\sum_{m=-\infty}^{\infty} f(m)\delta(m) = f(0)$.

The Fourier series is a representation of a periodic function $f(t)$ with period T, as a series of sines and cosines, or in the more general complex representation (using Euler's formula), a series of complex exponents:

$$f(t) = \sum_{n=-\infty}^{\infty} c_n exp\left[i\frac{2\pi n}{T}t\right], \tag{1.19}$$

With the Fourier coefficients: $c_n = \frac{1}{T}\int_{-T/2}^{T/2} f(t)exp\left[-i\frac{2\pi n}{T}t\right]dt$ where $n = 0, \pm 1, \pm 2, \ldots$

The 1-D Fourier transform, for the continuous case, is defined as:

$$F(\nu) = \int_{-\infty}^{\infty} f(t)exp[-i2\pi\nu t]dt, \tag{1.20}$$

And the inverse (almost symmetric) transform is:

$$f(t) = \int_{-\infty}^{\infty} F(\nu) exp[i2\pi\nu t] d\nu \tag{1.21}$$

Using Euler's formula: $exp(i\varphi) = cos\varphi + isin\varphi$, we may write the Fourier transform as:

$$F(\nu) = \int_{-\infty}^{\infty} f(t)[cos(2\pi\nu t) - isin(2\pi\nu t)]dt. \tag{1.22}$$

We may see that even if $f(t)$ is real, $F(\nu)$ can be complex. $F(\nu)$ is an expansion of $f(t)$ multiplied by sinusoidal terms, and it is only a function of the term ν, which determines the frequency of the sinusoidal functions. Thus the Fourier transform brings us to the frequency domain.

What is the Fourier domain equivalent of the convolution operation? We recall that the convolution between two continuous time functions $f(t)$ and $h(t)$ is $f(t) * h(t) = \int_{-\infty}^{\infty} f(\tau)h(t-\tau)d\tau$. Next we perform a Fourier transform on the convolution and use one of the Fourier properties: $\mathscr{F}\{h(t-\tau)\} = H(\nu)exp(-i2\pi\nu\tau)$ to obtain the convolution property: $f(t) * h(t) = F(\nu)H(\nu)$, so convolution in the time domain becomes multiplication in the frequency domain. The symmetrical property, called the modulation property, yields: $f(t)h(t) = F(\nu) * H(\nu)$.

Since we are dealing with sampled (discrete) images, we must address the sampling theorem. Sampling converts a continuous function into a sequence of discrete values. The sampled function can be written as:

$$f_s(t) = f(t) \sum_{n=-\infty}^{\infty} \delta(t - n\Delta T) \equiv f(t)s_{\Delta T} \tag{1.23}$$

We can insert $f(t)$ into the sum:

$$f_s(t) = \sum_{n=-\infty}^{\infty} f(t)\delta(t - n\Delta T) = \sum_{n=-\infty}^{\infty} f(n\Delta T)\delta(t - n\Delta T)$$
$$= \sum_{n=-\infty}^{\infty} f_n \delta(t - n\Delta T). \tag{1.24}$$

The Fourier transform of the sampled function is

$$\mathcal{F}\{f_s(t)\} = \mathcal{F}\{f(t)s_{\Delta T}\} = F(\nu) * S(\nu), \tag{1.25}$$

$$\text{Where } S(\nu) = \frac{1}{\Delta T} \sum_{n=-\infty}^{\infty} \delta(\nu - n/\Delta T), \tag{1.26}$$

Thus yielding, after some math, the shifting property:

$$F_s(\nu) = \frac{1}{\Delta T} \sum_{n=-\infty}^{\infty} F(\nu - n/\Delta T). \tag{1.27}$$

This means that if $f_s(t)$ is a sampled version of $f(t)$, then $F_s(\nu)$ is a periodic infinite sequence of copies of $F(\nu)$, with period $1/\Delta T$.

In order to avoid overlapping between different copies (usually referred to as aliasing), we need to satisfy the Nyquist criterion, stating that the sampling rate, $1/\Delta T$, must be at least twice the highest frequency present in $F(\nu)$. Sampling theorem suggests that if this criterion is met, then one may reconstruct the original $f(t)$ from its samples $f_s(t)$.

Since $F_s(\nu)$ is a continuous-frequency function (one we cannot present on the computer) we now ask what is the Discrete Fourier transform, or the sampled Fourier transform of the sampled function. The continuous transform of the sampled function is

$$F_s(\nu) = \int_{-\infty}^{\infty} f_s(t) exp[-i2\pi\nu t]dt, \tag{1.28}$$

But

$$f_s(t) = \sum_{n=-\infty}^{\infty} f(n\Delta T)\delta(t - n\Delta T) = \sum_{n=-\infty}^{\infty} f_n\delta(t - n\Delta T), \tag{1.29}$$

thus

$$F_s(\nu) = \sum_{n=-\infty}^{\infty} f_n exp[-i2\pi\nu n\Delta T], \tag{1.30}$$

Where $F_s(\nu)$ is continuous and infinitely periodic with period $1/\Delta T$.

Now we wish to sample $F_s(\nu)$, and keep its periodic nature; therefore the frequencies ν must be a rational multiplication of the period $1/\Delta T$

$$\nu = \frac{m}{M\Delta T}, \, for \, m = 0, 1, \ldots, M - 1. \tag{1.31}$$

This gives us

$$F_m = \sum_{n=0}^{M-1} f_n exp[-i2\pi mn/M], \, m = 1, 2, \ldots, M - 1, \tag{1.32}$$

Where the separation between samples in the frequency domain is $1/M\Delta T$.

The inverse Fourier transform is given by:

$$f_n = \frac{1}{M} \sum_{m=0}^{M-1} F_m exp\left[i2\pi mn/M\right], n = 1, 2, \ldots, M - 1. \tag{1.33}$$

The next step is to do the same for the 2-D case, and we begin by defining the 2-D Discrete Fourier transform as:

$$F_{k,l} = \sum_{m=0}^{M-1} \sum_{n=0}^{M-1} f_{m,n} exp[-i2\pi(mk/M + nl/N)], \begin{cases} k = 1, 2, \ldots, M - 1 \\ l = 1, 2, \ldots, N - 1 \end{cases}, \tag{1.34}$$

And the inverse 2-D transform is:

$$f_{m,n} = \frac{1}{MN} \sum_{k=0}^{M-1} \sum_{l=0}^{M-1} F_{k,l} exp\left[i2\pi(mk/M + nl/N)\right], \begin{cases} m = 1, 2, \ldots, M - 1 \\ n = 1, 2, \ldots, N - 1 \end{cases}, \tag{1.35}$$

Where $f_{m,n}$ is a digital image of size $M \times N$.

When comparing the spatial and frequency intervals, we can see that there's an inverse proportionality. If the image is sampled M times in the m direction (and N times in the n direction), using a step distance of ΔT, a total distance of $M\Delta T$ (and $N\Delta T$) is covered, which is related to the lowest frequency that can be measured:

$$\Delta k = \frac{1}{M\Delta T}, \Delta l = \frac{1}{N\Delta T} \tag{1.36}$$

We can easily see that both the 2-D Fourier transform and the 2-D Inverse Fourier transform are periodic in both directions. We could at this point go on and describe sharpening spatial filters, such as the Laplacian and Gaussian filters, or how IP deals with noise, but we prefer to focus now on the subjects at the heart of this book.

1.2 THE PROBLEMS OF SHRINKING FEATURE SIZE IN ULSI DEVELOPMENT AND FAILURE ANALYSIS

In a world in which technology develops very fast, new materials are introduced, and the dimensions of the elements continue to shrink, it is important that the techniques and the instruments used for failure analysis keep up with the pace of these changes. According to the 1999 International Technology Roadmap for Semiconductors (the 1999 ITRS), "The number and difficulty of the technical challenges continue to increase as technology moves forward" [6]. Quite a lot of research is done in the field of materials and processes, but little work is being done in the field of analysis. The 1999 ITRS contains many explicit references to characterization and analysis. Late in 1999, a subcommittee of the International Sematech Product Analysis Forum (PAF) reviewed the 1999 ITRS and identified a "top ten" list of challenges that the failure analysis community will face as present technologies are extended and future technologies are developed. Eight of the top ten are challenges of a technical nature; only two could be considered nontechnical in nature. Most of the challenges are common to several areas, from imaging small elements to fault simulation and modeling, from deprocessing to electrical defect isolation. Revolutionary changes require large multifaceted research efforts (unlike evolutionary changes that can be anticipated fairly easily).

A significant part of the time and effort spent in failure analysis is in locating where the defect causing the failure is placed on a die. This requires electrical measurements to make certain that the defect found completely explains the failure, whether it is a chip- or system-level failure. This is quite a significant challenge for a number of reasons. Since the die size increases while the feature size continues to become smaller, the defects become smaller with respect to the size of the total area. The signal to noise demands will also increase as the power supply (and on-chip signal) voltages decrease, which will cause larger sensitivity to interference from nearby elements. The current noninvasive localization and characterization techniques will not be sufficiently sensitive to determine whether or not a device is defective in the near future. Invasive techniques (e.g., photon beam probing and electron beam probing) may generate defects larger than the original defect causing the failure. All of these issues lead to the need for new techniques for failure analysis. Although there is some progress in the field, a lot of work has yet to be done [7–9].

Usually, finding the cause of a failure is done with optical microscopy or scanning electron microscopy (SEM). They have been used in a complementary manner where a defect was located optically, and then, once the location of the defect was known, it was inspected by SEM. In recent years, as the element sizes decreased, electron microscopy has become the more reliable tool and optical microscopy the less. However, a significant limitation of the SEM is that it is a near-surface imaging technique. Thus, it is harder to prepare samples for SEM, since the size of the defects become smaller, making it hard to separate the defect from the background. In X-ray tomography, on the other hand, one has to take the sample to the laboratory in order to detect and image the defects. This takes time, as does the fact that imaging is frequently an iterative process, which requires more time for higher resolutions.

In the near future many defects that are now visible will become invisible since the junctions and oxide gates on the chip will become much smaller. This will make imaging of such defects no longer possible and new techniques will have to be introduced. One example of this type of defect is tunneling in the gate oxide, leading to a failure quite different in nature from the current failures. Another challenge is that expensive, complex, and flexible test devices are required to simulate or reproduce the conditions leading to a failure, and sometimes the data becomes so massive that accessing the required portion of it is a very difficult task.

When referring to the simulation software, it is necessary to use the circuit schematics and physical design layout to identify where a defect is located on the chip or which part of the circuit caused the failure.

To summarize, the 1999 International Technology Roadmap for Semiconductors (the 1999 ITRS) provides a guide for the challenges facing the failure analysis community [7–9]. These challenges fall primarily into two categories: physical analysis and failure site isolation. The challenges in physical analysis are mainly due to the use of new materials and the decreasing size of element features. The failure site isolation challenges are mainly due to the increasing device complexity and the reduced accessibility to the circuits.

1.3 HIGH RESOLUTION IMAGING OF STRUCTURES

In this section in order to broaden the introductory exposure for readers, we will present an approach to metallic structure imaging that is not directly related to microelectronics. The readers should note that imaging of metallic and nonmetallic structures—and their various associated algorithms—has been widely studied in other fields, such as in medical imaging, and even in landmine detection [10,11].

In recent years, the transmission of light through metallic films with subwavelength structures has become a very extensive field of research. Effects such as wavelength filtering and transmission gain in such structures have been addressed in many research papers. Unlike visible wavelengths, for which subwavelength elements are difficult to analyze, long wavelengths may suggest a simpler way to investigate the elements. Bitzera and Waltherb [12] reported on a time-resolved terahertz imaging technique that allows for measuring of the electric field distribution in the near-field metallic samples. The authors demonstrated the propagation, attenuation, and interference of waves close to the surface of the sample. The work demonstrated terahertz pulse transmission through single-hole, 1-D, and 2-D arrays of subwavelength holes, as shown in Fig. 1.5(a).

The terahertz near-field imaging setup is based on the emission and detection of single cycle terahertz pulses by photoconductive antennas optically gated by the output of a mode-locked Ti:sapphire laser system. The emitted terahertz pulses are collimated and focused onto the front side of the sample by two off-axis parabolic mirrors. As shown in Fig. 1.5(b), the pulses propagate through the sample and the electric field after the sample is measured. The incident pulses are polarized horizontally and the detector orientation is such that it is sensitive to the horizontal component of the electric field. To create a complete image, the detector chip and the probe laser move together in x, y, and z directions relative to the sample. The spatial resolution was limited by the finite size of the probe beam focus and by the photoconductive gap between the detector electrodes and was on the order of 20 μm.

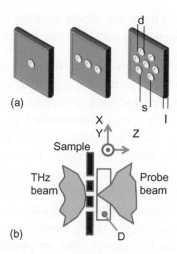

Figure 1.5 (a) Investigated metal samples with d = 300 μm, s = 450 μm, and thickness l = 300 μm. (b) Terahertz near-field imaging setup. The detector chip D can be shifted together with the probe laser beam in x, y, and z directions relative to the stationary sample.

The samples consist of 30 μm thick copper plates. Bitzera and Waltherb presented the electric field of a terahertz pulse transmitted through a single hole measured in a plane close to the back surface of the sample and along a cross section perpendicular to the sample surface through the center of the hole along the x-axis.

The authors observed the formation of a spherical wave that radiates from the hole into free space. The amplitude distribution along the surface was strongly anisotropic according to the surface wave propagation. Bitzera and Waltherb determined the amplitude spectra by a single spot measurement at the center of the hole's exit with the sample and obtained the spectra and the transmission efficiency. An enhancement of the transmitted amplitude has been reported before the study by Bitzera and Waltherb, and it is a basic aspect in the mechanism leading to extraordinary transmission through an array of holes. A model for analyzing the mechanism of passing through the hole involves the incident electric field E_{in}, which polarizes the surface by moving surface charges on the illuminated front side of the metal. Their movement is slowed down at the hole edges leading to a strong accumulation of charges, causing an oscillating dipole over the area of the hole. The interference between the incident wave and the radiation emitted from the dipole is constructive, and thus we obtain enhanced field transmission through the hole. When propagating backwards, the interference between the field radiated from the dipole and the field reflected from the neighboring metal is destructive, and thus the reflected beam is reduced.

Bitzera and Waltherb also investigated a two-dimensional subwavelength holes array consisting of seven hexagonally arranged holes. Under normal incidence, resonant coupling of an electromagnetic wave to a surface mode in a two-dimensional grating structure occurs when the wave vector of the surface wave matches the in-plane momentum provided by the holes array. This condition restricts the continuous spectrum of surface waves that arise from a single hole to a discrete spectrum. In their measurements, Bitzera and Waltherb observed enhanced surface mode amplitudes in a spectral window around the 0.7 THz frequency. Apart from decaying amplitudes with increasing distance from the center, the amplitude plot also depicted evident stationary minima between the holes.

1.4 FABRICATION TECHNIQUES IN ULSI INDUSTRY

A good summary of the fabrication techniques in the ULSI industry can be found in the work of Panwar [13], summing up most of the state-of-the-art data on fabrication techniques.

The main challenges in the field of nanofabrication relate to increasing fabrication reliability by optimizing throughput and minimizing process cost and complexity. Placing circuit patterns on semiconductor material gets more expensive as the integrated circuits (IC) become smaller. Lithography has the biggest impact on production costs of IC, and thus in this section we address the recent advances in lithography, mainly in the field of micro- and nanoelectronics.

According to Moore's law of doubling functionality on a chip every two years as the feature size on an integrated circuit decreases, fabricating the chips becomes increasingly difficult due to the complexity of the fabrication process (and the high cost). The fabrication of ultra large scale integrated (ULSI) devices requires the development of novel technologies for submicron processes. Nanostructure fabrication (also referred to as high-precision nanoscale lithography) is the key technology to the manufacturing of photonic components, chip-based sensors, biological applications, and more.

The first integrated circuits were manufactured in the early 1960s using optical lithography. Electron beam lithography was immediately used as a high-resolution alternative; however, optical lithography is still the most commonly used technique. In recent years, the cost of using optical technology has become quite high, giving way to new lithographic techniques, such as charged-particle lithography, uncharged-particle lithography, and nanoimprinting.

Let us start by explaining few basic terms from the field. Photolithography is a basic operation in the production of nanoelectronic devices (and also microelectronic devices). Lithography is a microfabrication technique used to make integrated circuits, microelectromechanical systems (MEMS), and nanoelectromechanical systems (NEMS). Lithography is a very detailed process used to transfer a pattern from a photomask to a layer. The main parts of a lithographic process are photo resist coating, exposure, development, and pattern transfer. The photo resist is a light-sensitive material used to form thin film, and it may be positive or negative.

Optical lithography is the technique for printing ultrasmall patterns onto silicon wafers. It is a method for patterning large areas with high throughput. The resolution of optical lithography can be determined by the wavelength of the imaging light and the numerical aperture of the projection lens. The resolution can be improved by decreasing depth of focus (DOF). Optical lithography is unmatched when considering the cost per pixel (one square unit of minimum resolution). A lithographic process capable of manufacturing state of the art chips must resolve the minimum feature size, overlay errors must be under tight restrictions, delicate complex patterns must be printed with high yield, and the overall cost of the process must be within reasonable limits.

193 nm lithography allows a large number of subelements within each pixel, and statistical fluctuations are quite small. Many 193 nm subelements are used to form the image, and thus shot noise effects are quite low. Optical lithography at ultraviolet (UV) wavelengths is the standard process for patterning 90 nm state-of-the-art devices in the semiconductor industry. Due to the high resolution, the intrinsic high throughput of optical lithography enables the development of many applications (not only for semiconductor electronics). UV optical lithography has developed to enable sub-100 nm resolution, and it has many applications in areas of nanotechnology beyond microelectronics applications (e.g., nanophotonics, MEMS, and NEMS). The transition

to 193 nm projection lithography has enabled mass production of microelectronic circuits with sub-100 nm dimensions.

Recently, the fabrication of the single electron transistor (SET) was done using optical lithography [14]. The optical lithography approach offers the possibility of integrating Si single-electron electronics with CMOS technology, and thus it is preferred to the electron-beam lithography for the fabrication of SETs. 93 nm immersion lithography has taken over as the leading technology for the 45 nm node on the International Technology Roadmap for Semiconductors (1999 ITRS), displacing 157 nm technology. The main benefit is the possibility of constructing projection optics with numerical apertures (NAs) > 1 by introducing an immersion fluid with a refractive index > 1. However, there are still several problems that must be addressed before the technology gains widespread acceptance.

Extreme ultraviolet (EUV) lithography works in reflection EUV mode and offers the prospect of simple design rules and simple optical proximity correction (OPC is a photolithography enhancement technique commonly used to compensate for image errors). Laser produced plasma (LPP) and discharged produced plasma (DPP) are the two main approaches used for EUV sources. EUV uses both reflective masks and mirrors for focusing, but 193 nm-based lithography is still better in terms of overall performance. NA of the lenses used in the process has increased from about 0.3 to 1.35 today with improvements in lens design and the use of immersion lithography. Simultaneously, the illumination wavelength has been reduced from 436 nm—about 20 years ago—to 193 nm for state-of-the-art scanners today [15]. The 22 nm technology node represents the last instance of using standard 1.35 NA immersion lithography based patterning for the critical layers. Recently, Wang et al. [16] presented Bragg gratings in Silicon-on-Insulator (SOI) strip waveguides fabricated by a single DUV (deep ultraviolet) lithography step [17].

The increasing cost and difficulty of creating masks make mask-based optical lithography too expensive when going beyond the 32 nm half pitch (HP) node. Here, electron-beam lithography (EBL) plays a complementary role to optical lithography. E-beam lithography offers higher resolution than optical lithography due to smaller wavelengths. E-beam lithography has the ability to fabricate patterns having nm feature sizes. In EBL, a finely focused electron beam is scanned over the substrate coated with a special electron-beam sensitive material called the resist. Here, electrons are emitted from the electron gun of a scanning electron microscope (SEM). Electron-beam direct-write (EBDW) is also referred to as maskless lithography.

EBL is an attractive technique for fabricating nanostructures. Recently, a fast and highly scalable room temperature nanomanufacturing process for the fabrication of metallic nanorods from nanoparticles was presented by Yilmaz et al. [18]. Applications for the produced 3-D nanorod arrays include sensors, CMOS Interconnects, and plasmonic-material-based sensors. Recently, Gonzalez-Velo et al. [19] suggested the use of EBL for localized microbeam irradiations. Localized irradiation of a unique device or a subcircuit can be performed automatically with EBL, and it can be used for performing localized irradiations of bipolar transistors. Maskless lithography provides an ultimate resolution without the risks of artifacts due to the usage of masks, but the productivity of the traditional single beam systems is quite low, making it very difficult for mass manufacturing. The main limitation is that it takes a very long time to expose an entire silicon wafer.

Several groups have proposed different multiple electron-beam maskless lithography (MEBML2) approaches, by using cell projections or by multiplying Gaussian beams to increase the throughput. Recently, The Mapper writing approach was suggested [20]. In it, the circuit layout in GDSII or OASIS format at a sub-nm addressing grid, whose file size can be up to hundreds of gigabytes, has to be pre-rasterized to a bit map writing format of a 3.5 nm grid. However, this is a slow process, not used in mass production.

X-ray lithography is the alternative process of electron-beam lithography, and it creates a submicrometer element size in a much cheaper way. X-ray lithography can choose from a large range of wavelengths, from about 0.4 to 100 nm. Synchrotron X-ray radiation is the most promising light source for transcribing superfine patterns of less than 0.2 μm. This capability, which is not possible for conventional photolithography, is required for 64 Mb DRAMs chips.

Dip-pen nanolithography is a direct-write method based on the AFM (Atomic Force Microscope). Dip-pen nanolithography is a type of scanning probe lithography that uses an AFM tip of a high resolution to write on a surface. In Dip-pen nanolithography (DPN), arbitrary nanoscale chemical patterns can be created by the diffusion of chemicals from the tip of an AFM probe to a surface. Dip-pen nanolithography has been demonstrated to create patterns on soft and hard materials on various substrates. An AFM tip is used as a pen to transfer ink materials onto a substrate, making it possible to fabricate small features from sub 100 nm to over a micrometer in size. DPN is a single-step process, which does not require the use of a resist. DPN has the ability to achieve the precise alignment of multiple patterns. Compared with nanografting, DPN does not require a preexisting monolayer and can be used when the deposited chemical must remain unconfined on the surface, such as in diffusion studies. Recently, a way to mass produce nanoscale features using parallel tip-arrays has been developed [21,22]. Multi-ink DPN has been studied to understand the patterning mechanism and to improve the resolution of multi-ink patterning [23–25]. Multi-ink DPN methodology can also be used in highly integrated protein patterning, and can provide the template for protein sensing and cell-binding studies.

Standard optical lithography is often used for microfabrication. However, single-photon absorption is limited to a 2-D process. The 3-D structures required for more complex devices are currently built up by consecutive lithographic steps. In order to advance technologically in areas such as microelectromechanical systems (MEMS) and nano-electromechanical systems (NEMS), 3-D optical data storage, and photonic crystals, 3-D structuring capabilities at the nanometer scale must be fully developed.

Two-photon lithography is a built-in 3-D lithography which has a high potential for constructing 3-D shapes arbitrarily in a single-step process. A femtosecond near-IR laser beam is focused into a resin, and subsequent photoinduced reactions, such as polymerization, occur in the close vicinity of the focal point, allowing the fabrication of a 3-D structure by directly writing 3-D patterns. Two-photon lithography can achieve much better spatial resolution than other 3-D microfabrication techniques. The most commonly applied two-photon technique is two-photon polymerization. Recently, a chemically amplified positive-tone two-photon system using a two-photon photo acid generator enabled the fabrication of buried 3-D structures [26,27].

The capability to selectively remove matter allows the efficient creation of small hollow features within a larger element. Positive-tone material systems like these provide the ability to pattern complex 3-D structures, such as waveguides and photonic lattices. Two-photon lithography is limited by low throughput due to the serial nature of the laser scanning process. One solution is to use parallel processing by means of dynamic diffractive optics.

Nanoimprint lithography (NIL) is a new nanolithography technique that has already demonstrated sub-10 nm resolution and high throughput, and it is inexpensive compared to optical lithography. Nanoimprinting can be used in various scientific disciplines, such as chemistry, biology, materials science. In recent years, imprint tools are being developed and improved all of the time, such as MII's "Imprio" series. The ITRS lithography roadmap includes the imprint lithography at the 32 and 22 nm nodes. One method for printing sub-100 nm geometries is step and flash imprint lithography (S-FIL), which—with respect to other imprinting processes—has the advantage that the template is transparent, and thus it may be used with conventional overlay techniques.

Nanoimprint lithography (NIL) avoids the low throughput limitation of EBL and is also the best method for fabricating nanometer scale patterns. The development of methods of nanolithography is crucial for the success of nanotechnology in mass production. Lugli et al. [28] discuss various versions of NIL, such as UV-NIL, room-temperature NIL (RTNIL), combined thermal and UV-nanoimprint lithography (TUV-NIL), and molecular beam epitaxy (MBE-RTNIL). Some other versions suggested by Nevludov et al. [29] are UV curing lithography (P-NIL-photocuring NIL), Hot embossing TNIL (thermal nanoimprint lithography), reversal imprint lithography, nanocontact printing, μ-contact printing (μCP), and nanotransfer printing (nTP). The highest resolution patterning demonstrations of sub-20 nm were performed using S-FIL by Willson [30]. The UV-nanoimprint replication process has many applications (e.g., microlens arrays and optical elements).

Imprint technology does not reduce dimensions, and thus the quartz template fabrication for monomer printing seems to be the most crucial part. The imprinting process generates a replica of the template feature onto the wafer. The quartz template determines the resolution, linearity, uniformity, and placement accuracy. Therefore, the performance and availability of specific templates affect the progress and the possible applications of nanoimprint technology. Using Gaussian beam pattern generators, one can realize the targeted resolution, but the throughput is not good and, in addition, present tools suffer from placement inaccuracy. On the other hand, variable shaped e-beam (VSB) writers provide the required throughput and placement accuracy, but most of them cannot meet the resolution target.

240 nm patterning using photocurable resin for electronic devices was suggested by Sakai [31]. Chou and Krauss [32] proposed roller-type nanoimprint lithography (RNIL). Thompson et al. [33] suggested the critical dimension uniformity and process latitude for 32 nm imprint masks. Ishizuka et al. [34] used sub-300 nm nanoimprint patterning to fabricate uniform gratings on composite semiconductors. Hirai et al. [35] analyzed high aspect ratio (of over 20) for nano imprint patterning with a width of 80 nm. Kehagias et al. [36] demonstrated submicron-level three-dimensional structures using nanoimprint lithography. Recently, a high accuracy, high precision, and cost-effective mastering process for wafer level camera image sensor applications using step-and-repeat nanoimprint lithography was discovered by Kreindl et al. [37].

Another approach is to use block copolymers. Block copolymer nano-fabrication can provide either objects with feature sizes on the order of tens of nanometers or large-area periodic functional structures. However, in many applications, a simple periodic structure is not good enough and there's a need for spatial control of the microdomains. Other lithographic methods have been used to form patterned substrates for the purpose of achieving spatial control over block copolymer nanostructures, besides photolithographically patterned substrates. E-beam lithography may result in a very good spatial control of functional microdomains, but this direct-write patterning process takes too much time for large-area integration of functional devices. This emphasizes the need for techniques for rapid patterning of functional nanostructures in real-time applications. Ober and coworkers have successfully developed a novel block copolymer system, the poly(a-methyl styrene)-block-poly(4-hydroxystyrene) system [38,39], to achieve spatial control through high-resolution deep UV lithographic processes, and submicron-sized patterns were generated through simple fabrication processes. Additionally, this block copolymer was automatically aligned with vertical orientations during spin coating over a wide range of film thicknesses (40 nm−1 mm), thereby avoiding tiresome alignment procedures. Stoykovich et al. [40] reported that ternary blends are capable of self-assembling on chemically patterned substrates to form periodic arrays.

To summarize, lithography is a key feature to advancing any technology. For electronic circuitry (and other fields), the smaller, the better, in terms of speed, power consumption, and cost. Many present and future lithographic technologies have been discussed in this chapter. Dimension in the microelectronic industry will continue to go down and this tends to create issues with particle detection, contamination control, and yield enhancement. Power consumption is also something to be taken into account in the provision of efficient, reliable, and cost-effective systems. As in most cases, progress in the field of lithography has been driven by industrial needs. Since industry needs constantly change, so does the technology, and while some technologies fail, others advance rapidly.

REFERENCES

[1] Van Pelt R, Bescós JO, Brecuwer M, Clough RE, Gröller ME, Romenij BT, et al. Interactive virtual probing of 4D MRI blood-flow. IEEE Trans Vis Comput Graph 2011;17 (12):2153−62.

[2] Gonzalez RC, Woods RE. Digital image processing. 3rd ed. Prentice Hall; 2007.

[3] Prewitt JMS. Object enhancement and extraction. Picture processing and psychopictorics. Academic Press; 1970.

[4] Sobel I, An isotropic $3 \times 3 \times 3$ volume gradient operator Technical report, Hewlett−Packard Laboratories, April 1995.

[5] Fourier JBJ. Théorie analytique de la chaleur. Cambridge Library Collection − Mathematics; 2009 (original publication 1822).

[6] Semiconductor Industry Association. International technology roadmap for semiconductors: 1999 edition. Austin, TX: International SEMATECH; 1999. p. iii.

[7] Joseph TW, Anderson RE, Gilfeather G, LeClaire C, Yim D. Semiconductor product analysis challenges based on the 1999 ITRS. AIP Conf Proc 2001;550:53−6.

[8] Wagner, LC. Failure analysis challenges. In: Proceedings of the eighth international symposium on the physical and failure analysis of integrated circuits, 2001.

[9] Kawata S, Inouye Y, Verma P. Plasmonics for near-field nano-imaging and superlensing. Nat Photonics 2009;3:388−94.

[10] Du J, Borden K, Diaz E, Bydder M, Bae W, Patil S, et al. Imaging of metallic implant using 3D Ultrashort echo time (3D UTE) pulse sequence. Proc Int Soc Magn Reson Med 2010;18:132.

[11] Du Bosq TW, Lopez-Alonso JM, Boreman GD, Muh D, Grantham J, Dillery D. Millimeter wave imaging system for the detection of non-metallic objects. In: Detection and

Remediation Technologies for Mines and Minelike Targets, Thomas Broach J., Harmon Russell S., Holloway John H. (Eds)., Proceedings of the SPIE, XI, Vol. 6217, 621723, 2006.

[12] Bitzera A, Waltherb M. Terahertz near-field imaging of metallic subwavelength holes and hole arrays. Appl Phys Lett 2008;92:231101.

[13] Panwar R. Recent developments, issues and challenges for lithography in ULSI fabrication. Int J Electron Comput Sci Eng (IJECSE) 2012;1(2):702–11.

[14] Yongshun Sun R, Singh N. Room-Temperature Operation of silicon single electron transistor fabricated using optical lithography. IEEE Trans Nanotechnol 2011;10(1):96–8.

[15] Sivakumar S. EUV lithography: prospects and challenges design automation conference (ASP-DAC), 16th Asia and South Pacific, 2011, 402 pp.

[16] Wang X, Shi W, Vafaei R, Jaeger NAF, Chrosstowski L. Silicon-on-insulator Bragg grating fabricated by deep UV lithography. IEEE Proceedings of the Communications and Photonics Conference and Exhibition (ACP), Asia, 2010, pp. 501–502.

[17] Giuntoni I, Stolarek D, Richter H, Marschmeyer S, Bauer J, Gajda A, et al. Deep-UV Technology for the fabrication of Bragg gratings on SOI rib waveguides. IEEE Photon Technol Lett 2009;21(24):1894–6.

[18] Yilmaz C, Kim T-H, Somu S, Busnaina AA. Large scale nanorods nanomanufacturing by electric field directed assembly for nanoscale device applications. IEEE Trans Nanotechnol 2010;9:653–8.

[19] Gonzalez-Velo Y, Boch J, Pichot F, Mekki J, Roche NJH, Perez S, et al. The use of Electron-Beam Lithography for localized microbeam irradiations. IEEE Transa Nucl Sci 2011;58(3):1104–11.

[20] Chen JJH, Krecinic F, Chen J-H, Chen RPS, Lin BJ. Future electron-beam lithography and implications on design and CAD tools. In: IEEE proceedings of the Design Automation Conference (ASP-DAC), 16th Asia and South Pacific, 2011, pp. 403–404.

[21] Salaita K, Lee SW, Wang X, Huang L, Dellinger TM, Liu C, et al. Sub-100 nm, centimeter-scale, parallel dip-pen nanolithography. Small 2005;1(10):940–5.

[22] Salaita K, Wang Y, Fragala J, Vega RA, Liu C, Mirkin CA. Massively parallel dip-pen nanolithography with 55000-pen two dimensional arrays. Angew Chem Int Edit 2006;45 (43):7220–3.

[23] Hampton JR, Dameron AA, Weiss PS. Double-ink dip-pen nanolithography studies elucidate molecular transport. J Am Chem Soc 2006;128(5):1648–53.

[24] Nafday OA, Haaheim JR, Villagran F. Site-specific dual ink dip pen nanolithography (TM). Scanning 2009;31(3):122–6.

[25] Nafday OA, Weeks BL. Relative humidity effects in dip pen nanolithography of alkanethiol mixtures. Langmuir 2006;22(26):10912–4.

[26] Zhou W, Kuebler SM, Braun KL, Yu T, Cammack JK, Ober CK, et al. An efficient two-photon generated photoacid applied to positive-tone 3D microfabrication. Science 2002;296 (5570):1106–9.

[27] Yu T, Ober CK, Kuebler SM, Zhou W, Marder SR, Perry JW. Chemically amplified positive resists for two-photon three-dimensional Microfabrication. Adv Mater 2003;15 (6):517–21.

[28] Lugli P, Harrer S, Strobel S, Brunetti F, Scarpa G, Tornow M, et al. Advances in nanoimprint lithography. IEEE Proc Seventh IEEE Conf Nanotechnol 2007;1179–84.

[29] Nevludov ISH, Palagin VA, Frizuk EA. Nanolithography and nanoimprinting, International Workshop on Optoelectronic Physics and Technology, OPT '07, 2007, pp. 63–67.

[30] Willson CG. A decade of step and flash imprint lithography. J Photopolym Sci Technol 2009;22:147−53.

[31] Sakai N. Photo-curable resin for UV-nanoimprint technology. J Photopolym Sci Technol 2009;22:133−45.

[32] Chou SY, Krauss PR. Imprint lithography with sub-10 nm feature size and high throughput. Microelectronic Eng 1997;35:237−40.

[33] Thompson E, Selinidis K, Maltabes JG, Resnick DJ, Sreenivasan SV. Process control for 32 nm imprints masks using variable shape beam pattern generators. Microelectron Eng 2009;86:709−13.

[34] Ishizuka S, Nakao M, Mashiko S, Mizuno J, Shoji S. Fabrication of uniform gratings on composite semiconductors using UV nanoimprint lithography. J Photopolym Sci Technol 2009;22:213−7.

[35] Hirai Y, Yoshida S, Takagi N, et al. High aspect pattern fabrication by nano-imprint lithography using fine diamond mold. Jpn J Appl Phys 2003;43:3863−6.

[36] Kehagias N, Reboud V, Chansin G, et al. Submicron three-dimensional structures fabricated by reverse contact UV nanoimprint lithography. J Vac Sci Technol 2006;B24:3002−5.

[37] Kreindl G, Glinsner T, Fodisch R, Treiblmayr D, Miller R. IEEE Transactions 2010:347−51.

[38] Ober CK, Li M, Douki K, Goto K, Li X. Lithographic patterning with block copolymers. J Photopolym Sci Technol 2003;16:347−50.

[39] Du P, Li M, Douki K, Li X, Garcia CBW, Jain A, et al. Additive-driven phase selective chemistry in block copolymer thin films: the convergence of top-down and bottom-up approaches. Adv Mater 2004;16(12):953−7.

[40] Stoykovich MP, Müller M, Kim SO, Solak HH, Edwards EW, de Pablo JJ, et al. Directed assembly of block copolymer blends into nonregular device-oriented structures. Science 2005;308(5727):1442−6.

CHAPTER 2

New Image Processing Methods for Advanced Metallization in Micro- and Nano-Electronics

2.1 CHARACTERISTICS OF METAL ULTRATHIN FILMS' MICROSTRUCTURES

The phenomenal success in microelectronics within the last few decades has revolutionized the world of electronics and dramatically changed humans' daily lives. Different industrial segments, such as health care, internet, aviation, automotive, telecommunication, and information cannot substantially advance without this multidisciplinary technology. The semiconductor microchips that lie at the heart of microelectronics are utilized in almost all electronic equipment in use today, starting with toys and electronic games, and ending with sophisticated computers, mobile phones, and satellites. In the 21st century, the growing population, longer life expectancies. and a constantly improving quality of life are dictating that electronic equipment become more powerful, functional, and cost-efficient. Therefore, modern chips have to integrate an ever-increasing number of functional units, each of which is built from active and passive elements. As a result, the number of active, as well as passive, elements on an individual microprocessor chip is skyrocketing and has already exceeded a few hundred million.

To implement assigned functions, each active element requires a power supply and ground voltage, which are delivered through metal interconnection lines. The interconnection lines, which are also utilized

New Approaches to Image Processing based Failure Analysis of Nano-Scale ULSI Devices.
DOI: http://dx.doi.org/10.1016/B978-0-323-24143-4.00002-1

to transfer data from one logic cell to another, are an important and integral part of each chip. In state-of-the-art devices, interconnects constitute cumbersome and branching multilevel networks. The metal levels (up to 12) are interconnected through vias and separated by dielectric layers. The passive elements—for instance, interdigital capacitors—are also built from metal lines.

The ever-increasing performance of microchips is one of the characteristics of microelectronics. The switching time decreases with transistors' downsizing. Currently, leading-edge devices are manufactured at the 32 nm CMOS technology node [1]. However, the overall operating speed is also defined by the signal delay across the interconnects as a result of their resistive-capacitive load, which increases due to the growing interconnects' resistance caused by their downsizing and because of denser packing of the interconnects. Moreover, increasing resistance results in signal rise/fall time degradation. This not only limits the operation frequency, but also increases power dissipation, decreases reliability, and causes logic faults. Hence, to make faster devices, scaling just the transistors and optimizing the circuit design is no longer sufficient.

The key to increasing performance and maintaining high device operation speed lies in optimizing the interconnect system (e.g., by introducing new materials and processes). This has already been expressed by the transition from aluminum to copper interconnects at around the 180 nm CMOS technology node and, more recently, by the introduction of novel insulators with low dielectric constants. The adoption of copper helped mitigate deleterious effects pertaining to resistance, corrosion, and electromigration (EM) of the aluminum interconnects [2]. At the same time, it was a net creator of additional, not fewer, intractable challenges for the back end of line (BEOL) R&D people.

Copper naturally has a higher propensity to diffuse into dielectrics than aluminum does. Therefore, to prevent copper migration—which hurts device functionality and reliability—diffusion barrier layers that encapsulate the copper interconnect were introduced [3]. Usually, they are compounds of tantalum or titanium. Another problem with copper is that it exhibits weak adhesion to many other materials. Thus, an adhesion-enhancing layer (liner) on top of the diffusion barrier is essential [4]. Furthermore, because of the lack of volatile copper compounds, copper could not be patterned by the previous techniques of photoresist masking and plasma etching that had been successfully used with aluminum. Thus, IBM has developed the dual damascene process in which vias and trenches for copper interconnects are first simultaneously etched directly in an insulator. Then liners and copper layers are deposited to fill in the pattern [5]. In this technology, several levels of copper metallization are materialized by using a planarization technique called chemical–mechanical polishing (CMP) and electro-chemically deposited (ECD) copper. However, the ECD-Cu process used to fill trenches and vias requires the deposition of a copper seed layer as a highly conductive electrode and to improve plated copper adhesion, microstructure, and EM characteristics [6].

Even though the dual damascene processing on 300 mm wafers has firmly established its position in the *IC* industry, the major reliability challenges that must be overcome in state-of-the-art devices are at the metallization processing level (i.e., at the BEOL level). The three most critical process elements affecting copper interconnect reliability are: copper vias (majority of early reliability failures are attributed to vias), copper/liner/barrier and barrier/insulator interfaces, and liner coverage [7]. A good barrier, liner, and seed coverage in the trenches is a prerequisite for good copper fill. By appropriate liner selection and conformal deposition, copper diffusion along the copper/liner interface is minimized. A good adhesion can also prevent the delamination between the layers, and can help to reduce the possibility of void formation within and/or underneath the vias [8].

With smaller feature sizes, the conformal and defect-free deposition of the ever-shrinking liners and diffusion barriers becomes more and more difficult. In particular, the realization of a homogeneous liner for the via holes and trenches of the interconnect lines is a challenge because of the high aspect ratio (AR) in this location [9]. The traditionally employed physical vapor deposition (PVD) technique is reaching its limits for depositing conformal thin films as the AR of submicron features increases, because of its intrinsic directional (i.e. anisotropic) film growth. Even the improved ionized PVD (iPVD) is facing major difficulties already with an AR of 3:1, whereas according to the ITRS roadmap, the AR will continue growing. The diffusion barrier, liner, and copper seed layer integration is also very challenging due to many complex process requirements.

The copper diffusion barrier process must fulfill several compatibility requirements without which the integration may fail [10]. The situation has just been exacerbated by the introduction of insulators with low dielectric constants (low-k dielectrics) to reduce the fast-growing capacitance due to the decreasing space between adjacent interconnect lines. If a conventional diffusion barrier deposited by PVD technique is used, porous sidewalls of vias and trenches make the barrier films discontinuous [11]. Because of nonconformal growth of PVD Cu, there is overhang growth on the top edge of the vias. Thicker film on the edge causes locally higher current density in the ECD process, resulting in a higher growth rate at this spot, and eventually leaving a void in the middle of the via. The diffusion barriers also appreciably contribute to interconnect resistance, and the contribution is proportional to the barriers' thickness [12]. Therefore, to improve devices' performance, major efforts are currently under way in two directions: the study of new ultrathin film deposition methods free of the shortcomings pertaining to PVD and the quest for new materials capable of acting as a robust liner layer with intrinsic copper direct-plate and adhesion characteristics (i.e., no need for an additional copper seed layer) [1]. The latter issue is expected to be addressed by employing ruthenium or cobalt. Ruthenium has a low solid solubility in copper up to 900°C, and no new phases have been detected at the ruthenium−copper interface after annealing at 800°C, thus indicating stability of the interface [13]. Cobalt has also demonstrated performance as a liner and adhesion promoter between a diffusion barrier and copper [14].

In the metallization field of micro- and nanoelectronic devices, the ITRS roadmap points to atomic layer deposition (ALD)—a method relying on alternate pulsing of the precursor gases and vapors onto the substrate surface, subsequent chemisorption, and surface reactions of the precursors [15]—as the most promising alternative to PVD. ALD actually represents one of the most significant developments for depositing uniform and conformal thin films with accurate thickness control and homogeneity on large-area substrates and three-dimensional nanostructures. The ability to achieve a stable growth (thickness increase at a constant rate in each deposition circle) and a straightforward growth of different multilayer structures makes ALD a very attractive and promising method for addressing a wide range of even very subtle requirements.

Another promising candidate is Electroless Deposition (ELD), which offers high-quality ultrathin films that, in their turn, meet requirements for the sub-45 nm ULSI interconnects, contacts, and via contacts, as well as in high AR structures for micro- and nanoelectronic mechanical systems [16]. In addition, ELD is a low-cost, relatively simple, and highly selective method, which potentially allows the elimination of some lithography and chemical mechanical polishing steps. It can also be produced on either conductive or dielectric substrates. In the latter case, however, surface activation before ELD is required. This has a strong influence on the deposition kinetics and properties of the final metal films.

The metal ultrathin films are not only employed in microchips, but also in many other microelectronics devices. The discovery of the giant magnetoresistance (GMR) phenomenon in 1988 by Peter Grünberg in Fe/Cr/Fe sandwiches [17] and Albert Fert in Fe/Cr multilayers grown by molecular beam epitaxy (MBE) [18] has launched spintronics, as a new field of microelectronics. In the early 1990s it was reported that Co/Cr, Co/Ru, and Co/Cu multilayers prepared by sputtering, demonstrated the same GMR phenomenon [19]. Up until now, there have been reports on the GMR characteristics of deposited Co−Cu/Cu, Ni−Cu/Cu, Co−Ag/Ag, Co−Au/Au, Co−Ru/Ru, Co−Ni−Cu/Cu, Fe−Ni−Cu/Cu, Fe−Co−Cu/Cu, Co(−Cu)−Zn/Cu and Fe−Co−Ni−Cu/Cu multilayer films [20]. Today, devices based on such multilayer structures are for the most part manufactured via physical vapor deposition (PVD) methods, mainly sputtering. It has been demonstrated that electrodeposition (ED) is also capable of producing multilayered magnetic nanostructures exhibiting a GMR effect.

The GMR phenomenon may not only be exploited to satisfy the never-ending and ever-increasing demand for efficient data storage and the sensing of weaker magnetic fields, but also for the detection, analysis, and characterization of magnetic nanoparticles attached to biological mediums such as antibodies and DNA [21]. European research groups were the first to show a spin valve (SV)-based magnetic sensor [22]. A research group at Stanford University fabricated a magnetic biochip with nanoparticle tags based on a SV GMR array [23]. GMR biosensors have demonstrated increased sensitivity and higher performance than fluorescent labels. Stanford scientists are developing a microchip device that is based on the GMR phenomenon to detect cancer protein markers in blood [24]. One of the biggest challenges is that all GMR-based devices require precision deposition tools to control the film quality. The films used in the GMR sensors are very thin and the tolerances are tight.

The excellent optical properties of silver also make it a suitable material for various applications in optics and electro-optics. In near-infrared (NIR) wavelengths, the reflectance is very high; then it decreases to the visible region, and then drops considerably into the ultra-violet one [25]. For instance, silver is widely used as guiding layers in hollow waveguides for IR transitions [26]. Silver also becomes increasingly attractive for ULSI applications as it interconnects and contacts with predicted features shrinking down to values below 50 nm. The quality of ultrathin silver films strongly depends on the deposition technique.

From the above it follows that on the one hand, robust and reliable metallic ultrathin films with excellent electrical and optical properties are essential for further advances in the field of micro- and nanotechnology. On the other hand, it is becoming more and more difficult to realize the adequate growth of these films. The successful development of novel metal thin films' deposition techniques, as well as the constant improvement of already-existing methods, strongly depend on the availability of reliable characterization tools, such as secondary ion mass spectroscopy (SIMS), X-ray photoelectron spectroscopy (XPS), high-resolution scanning electron microscopy (HRSEM), and several others. These enable the examination of the nucleation effects, the growth-rate of the films, and the different chemical—physical phases during their growth processes.

The theory of metallic ultrathin films suggests that—independent of deposition techniques—when metallic thin films grow on insulating surfaces (which is very often the case in modern micro- and nanoelectronics) they pass through a sequence of morphological changes [25]. In the early stages, in which the film is extremely thin, it consists of isolated compact islands. As the deposition proceeds, these islands grow and become larger, but still remain compact islands. At some certain film thickness, these islands coalesce into near-equilibrium compact shapes forming percolating structures. Finally, the channels between the structures are filled in and a continuous, free of holes, film is created. These three growth stages are demonstrated in Figs. 2.1, 2.2 and 2.3, which are SEM images, obtained during silver and copper ELD.

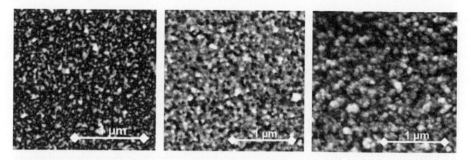

Figure 2.1 HRSEM images of Ag films grown using the ELD method on SiO₂/Si surfaces activated by Pd. With increasing thickness, three distinct stages can be observed: compact and isolated islands, coalescing equilibrium shapes, and continuous hole-free film.

Figure 2.2 HRSEM images of Cu films grown using the ELD method on SiO₂/Si surfaces activated by Pd. With increasing thickness three distinct stages can be observed: compact and isolated islands, coalescing equilibrium shapes, and continuous hole-free film.

Figure 2.3 HRSEM images of Ag films grown using the ELD method on SiO₂/Si surfaces activated by 5 nm AuNPs for 10 min. With increasing thickness three distinct stages can be observed: compact and isolated islands, coalescing equilibrium shapes, and continuous hole-free film.

On the other hand, it can be clearly seen that growth rates and tendencies are considerably different from metal to metal. For instance, copper, at the nucleation stages, tends to grow laterally, whereas silver tends to yield a 3-D granular growth. Furthermore, for the same material, the geometry of the isolated compact islands could be different depending on the substrate properties on which the materials are deposited and the deposition technique itself. In the case of ELD, in which the activation of the insulating surface by the catalyst is crucial for the following metallization, the deposition's dynamics are strongly influenced by the selected catalyst. An aqueous palladium (Pd) colloid [27] or Pd ionic solution [28] as well as colloid solutions of different sizes (e.g. 5 nm, 15 nm, and 25 nm) of gold nanopartricles (AuNPs) [29], have been successfully used as catalysts for metallic ultrathin film deposition on dielectrics' surfaces. In the case of AuNPs, the size of the nanoparticles and the activation time also substantially define the dynamics' growth [30].

The trends described in this subchapter clearly emphasize the critical role of high-quality SEM images in the investigation of the growth properties of different ultrathin metallic films employed in micro- and nanoelectronics.

2.2 INCREASED PRODUCTIVITY BY OBVIATING STEPS OF SELECTION OF MEASUREMENT CONDITIONS

2.2.1 Introduction

The rapid development of ULSI microchips, accompanied by a continuous downscaling, opens up new opportunities. At the same time, it produces intractable challenges for developers of microchips in general and for networks of metal interconnection lines (a critical part of each microchip) in particular. Today, most early failures of metal interconnects, which damage chip functionality, are attributed to the processing phase. The key to detecting these failures lies in the ability to obtain high-quality images of ultrathin film microstructure and surface coverage. Equally important, it is vital to obtain clear images of the filling of trenches and vias in multilevel interconnect structures, a very challenging processing phase due to the high aspect ratio (AR) in this location. However, the receipt of distinct scanning electron microscope (SEM) or transmission electron microscope (TEM) images of metallic ultrathin films, allowing for accurate estimation of their optical, mechanical, and electrical properties, is a nontrivial and quite subtle task, since they are affected by the poor conductivity of these nanoscale films (e.g., electron charging). The low-beam current required to enhance image quality results in rather tedious and time-consuming measurements.

At present, super-resolution and image enhancement methods are commonly employed to improve distorted images by reconstructing and enhancing high spatial frequencies. The resolution improvement is generally obtained by exploiting *a priori* knowledge about the inspected sample or the imaging system [31]. In addition, tools, such as histogram equalization, Fourier filtering, morphological operations, linear geometrical operators, and Sobel differential or Laplacian-based edge detection have been reported in the literature [32].

As opposed to optical microscopy (which usually refers to light in the UV–visible–NIR range), SEM and TEM imaging relates to electronic optics, and thus different *a priori* considerations are related to the spatial blurring. This serves as the basis for creating novel image enhancement algorithms in which the methods that have been successfully used for light imaging systems are adapted to the specific problems involved when SEM- and TEM-based inspection is applied to metal thin film microstructures. Such an image improvement technique has been developed and is described in the following section.

2.2.2 The Novel Algorithm

The proposed algorithm takes into account *a priori* knowledge about microstructure grain size and shape ranges. In the image enhancement process, we apply an image transformation, called the K-factor transform, which has been shown to be well-suited for dealing with shadowed images [33,34]. This transformation decomposes a given image into a product of several binary images. The K-factor decomposition of an image $I(x,y)$ is defined by:

$$I(x,y) = \prod_{n=1}^{N} f_n = \prod_{n=1}^{N} \frac{1 + k^n g_n}{1 + k^n}, \tag{2.1}$$

where k is a threshold less than 1, f_n is the K-factor harmonics, (x,y) are spatial coordinates and g_n is a binary image computed by:

$$g_n(x,y) = \begin{cases} 1 & \dfrac{I(x,y)}{\prod_{j=1}^{n-1} f_j(x,y)} > \dfrac{1}{1+k^n}, \text{ and} \\ 0 & elsewhere \end{cases} \tag{2.2}$$

$g_1(x,y)$ is determined using:

$$g_1(x,y) = \begin{cases} 1 & I(x,y) > \dfrac{1}{1+k}, \\ 0 & elsewhere \end{cases} \tag{2.3}$$

where N denotes the number of harmonics. When used in image reconstruction, $N = 8$ is sufficient to obtain a reconstruction with negligible visible error.

In each one of the decomposed binary images, the location of a grain is determined using a correlation operator. The correlation is performed with a circle having a diameter matching the anticipated range of grain diameter. The correlation $C_{f_n W}$ between K-factor harmonic f_n and an image of a circle (denoted as W) is defined by:

$$C_{f_n W}(x, y) = \iint f_n(x', y') W^*(x' - x, y' - y) dx' dy' \tag{2.4}$$

The resulting 2-D correlation image is given a threshold of 85% of the peak correlation value, so that around each identified grain a circle will appear (the result of giving a correlation peak a threshold). Each decomposed image is processed independently, following which the inverse K-factor transform is used in order to reconstruct the gray-level image out of the enhanced decomposed harmonics. The visual example of the algorithm application is presented in Fig. 2.4.

For very flat samples, such as extremely homogeneous layers, it is better to correlate between the image of a circle and the differentiated image of the K factor harmonic f_n. This is preferable since in this case the correlation is very uniform, and all of the image will pass the threshold criterion. Hence the 2-D correlation image with the differential image will have a lower threshold at 10%.

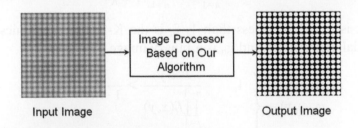

Input Image   Output Image

➤ The input image is blurred, and circles can hardly be discerned.

➤ The circles' dimensions are known.

➤ Based on *a priori* knowledge, the image was reconstructed, and the circles are clearly seen in the output image.

Figure 2.4 General view of the proposed algorithmic approach.

The result is the original image with enhancement obtained at locations of the low contrast grains. This enhanced image was then given a threshold at 10% of the gray level range (i.e., the upper 90% gray levels were accepted as objects) in order to determine the surface coverage.

2.3 DEMONSTRATION OF METHOD CAPABILITIES

The practicability of the proposed algorithm has been examined by applying it to blurred images of copper and silver ultrathin films deposited by an electroless plating method (a short description of this deposition method can be found in the previous section).

Figure 2.5(a) shows an HRSEM image of a ~ 60 nm thick silver film deposited on a SiO_2 substrate and vacuum annealed at 350°C. In this fuzzy image, it is hardly possible to distinguish grains, let alone to determine their sizes and surface coverage. For the sake of comparison, Fig. 2.5(b) shows the image following histogram equalization, a common tool to improve distorted images. However, even after applying this tool, the grains are still difficult to resolve. Following application of our algorithm, the image quality is substantially improved, as can be clearly seen in Fig. 2.5(c). In the improved image, the grains in the range of 35–80 nm are easily resolved. It is possible to see that some grains create mounds (clusters). The processed image reveals that the film is not continuous, as grooves (pores) are conspicuously discerned. Our MATLAB-based calculations show that the coverage in the original blurred image is 99.8%, and the coverage in Fig. 2.5(b) is 98.7%, whereas in the processed image by the proposed algorithm it stands at only $\sim 91.3\%$. This indicates that the potential for error in this critical parameter is alarming.

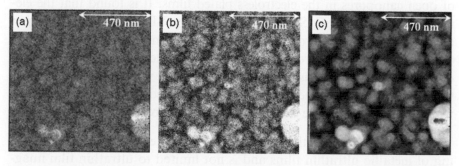

Figure 2.5 (a) The HRSEM image of a ~60 nm thickness silver film deposited by electroless plating on SiO_2 substrate and annealed at 350°C. The same image after it was processed by a typical currently available image improvement tool (b) and developed by the discussed algorithm (c).

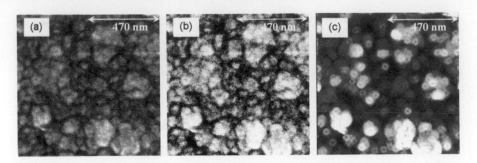

Figure 2.6 (a) The HRSEM image of a ~55 nm thickness copper film deposited by electroless plating on SiO₂ substrate. The same image after it was processed by a typical currently available image improvement tool (b) and the image obtained after being processed by the discussed algorithm (c).

Images of ~55 nm copper thin films were also significantly improved, as can be seen in Fig. 2.6(c), after comparing it with the original image, demonstrated in Fig. 2.6(a), and the image obtained following histogram equalization, demonstrated in Fig. 2.6(b). Using the algorithm, it was also found that this film is not continuous, as grooves (pores) are clearly seen. The calculations show that the surface coverage is only 93.5% and not 97.5% as it could have been wrongly concluded from the original blurred image.

It should be noted that, in our previous studies under the same experimental conditions on both films, the surface coverage was found to be lower than 95% [35,36], which is consistent with the results obtained from images processed with our algorithm. Furthermore, the extent of surface coverage obtained from the improved microscopy image correlates well with an X-ray photoelectron spectroscopy (XPS) analysis of the thin-film, summarized in Table 2.1. The XPS analysis clearly reveals the presence of open nanopores in the electroless plated films. It is found that the elemental signals of N, C, Si, and O recorded by XPS through the pores in the Cu layer originate from a self-assembled monolayer (SAM) of silane (used as an adhesion promoter). In addition, the electrical conductivity of the films, measured using a four-point probe, was larger than that of Cu-films having the same thickness, but deposited by E-beam technique, which is explained by the existence of the film porosity.

The presented algorithm can be easily adjusted for future downscaling of metallic ultrathin films and is not limited to ultrathin film imaging. It may also be used to substantially improve the images of trench and via filling, a very active and hot issue in nano- and microelectronic

Table 2.1 The XPS Depth Profile Data for 55 nm Cu Electroless Plated Film

Sputtering Time, t (min)	Binding Energy, eV			Atomic Concentration, at. %					
	O1s	Si2p,	N1s	N	O	C	Si	Au	Cu
0				0.7	25.7	30	nd[a]	nd	42
2.0				1.0	1.0	0.9	0.7	nd	96.3
7.0	532 + small band at ≈ 529.5		399 + band at ≈401.5[c]	1.0	1.0	1.1	nd	0.9	96
7.5		≈ 103 + band at ≈ 101−101.5[b]		0.9	1.2	0.4	1.6	0.9	95
11.5	534.5 + band at ≈ 532	≈ 104 + band at ≈ 102[b]		1.0	13.6	0.4	4.6	1.4	79
13.0	534.4	104.5		0.4	22.4	0.9	11.1	1.1	64

[a]nd—not detected.
[b]The low-energy shoulder (∼101.7 eV) to the Si2p peak from SiO_2 (103−104 eV) is considered as a signature of silyl−alkyl moieties, that is $[-(CH_2)_3-Si\equiv]$ [10].
[c]The N1s signal at 399 eV and high energy shoulder at 401−401.5 eV were assigned to the nitrogen in the amino-groups [37].

technology. The practicability of the algorithm for these applications has been examined on an original HRSEM image of a trench (filled with silver deposited by electroless plating), shown in Fig. 2.7(a) and compared with that processed by histogram equalization, shown in Fig. 2.7(b). Both of these images do not satisfactorily enable us to define whether the voids—a severe reliability problem—exist. Following processing of the original image with the proposed algorithm, as the image demonstrates in Fig. 2.7(c), one can see that the filled trench is free of voids.

To sum up, in this subchapter a novel algorithm—different from existing image enhancement methods, which are based on *a priori* knowledge about the inspected sample or the imaging system—was introduced. It takes into account *a priori* knowledge of the microstructures' grain size and shape ranges. The algorithm's ability to improve the SEM and TEM images of ultrathin film microstructures, as well as of via and trench filling has been presented. The algorithm was tested on blurred HRSEM images of silver and copper ultrathin films, which have been deposited by electroless plating, as well as on silver filled trenches. The processed images indicate that the algorithm has the potential to assist researchers in the time-consuming process of obtaining clear SEM and/or TEM images of microelectronic chips.

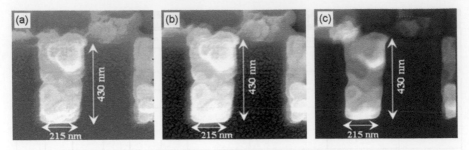

Figure 2.7 The HRSEM images of trench filled with Ag deposited by electroless plating: original (a), by a typical currently available image improvement tool (b), and by the discussed algorithm (c).

REFERENCES

[1] International Technology Roadmap for Semiconductors. 2011 Edition, Interconnect.

[2] Andricacos PC. Copper on-chip interconnections. Electrochem Soc Interface 1999;8:32−7.

[3] Havemann RH, Hutchby JA. High-performance interconnects: an integration overview. Proc. IEEE 2001;89(5):586−601.

[4] Lane MW, Murray CE, McFeely FR, Vereecken PM, Rosenberg R. Liner materials for direct electrodeposition of Cu. Appl Phys Lett 2003;83:2330−2.

[5] Andricacos PC, Robertson N. Future directions in electroplated materials for thin-film recording heads. IBM J Res Dev 1998;42:671−80.

[6] Andricacos PC. Copper on-chip interconnections. Electrochem Soc Interface 1999;2:32−7.

[7] Li B, Sullivan TD, Lee TC, Badami D. Reliability challenges for copper interconnects. Microelectron Reliab 2004;44:365−80.

[8] Lane WM, Liniger E, Lloyd JR. Relationship between interfacial adhesion and electromigration in Cu metallization. J Appl Phys 2003;93(3):1417−21.

[9] Fischer AH, Aubel O, Gill J, Lee TC, Li B, Christiansen C, et al. Reliability challenges in copper metallizations arising with the PVD resputter liner engineering for 65 nm and beyond. IEEE IRPS Proc 2007;511−15.

[10] Elers K-E, Blomberg T, Peussa M, Aitchison B, Hauka S, Marcus S. Film uniformity in atomic layer deposition. Chem Vap Deposition 2006;12(1):13−24.

[11] Roule A, Amuntencei M, Deronzier E. Seed layer enhancement by electrochemical deposition: the copper seed solution for beyond 45 nm. Microelectron Eng 2007;84:2610−14.

[12] Kumar S, Greenslit D, Chakraborty T, Eisenbraun ET. Atomic layer deposition growth of a novel mixed-phase barrier for seedless copper electroplating applications. J Vac Sci Technol A 2009;27(3):572−6.

[13] Chyan O, Arunagiri TN, Ponnuswamy T. Electro-deposition of copper thin film on ruthenium. A potential diffusion barrier for Cu interconnects. J Electrochem Soc 2003;150: C347−50.

[14] Yang C, Cohen S, Shaw T, Wang P-C, Nogami T, Edelstein D. Characterization of ultrathin-Cu/Ru(Ta)/TaN liner stack for copper interconnects. IEEE Electron Device Lett 2010;31:722−4.

[15] Leskelä M, Ritala M. Atomic layer deposition (ALD): from precursors to thin film structures. Thin Solid Films 2002;409:138−46.

[16] Warrender JM, Aziz MJ. Evolution of Ag nanocrystal films grown by pulsed laser deposition. Appl Phys A: Mater Sci Process 2004;79:713−16.

[17] Binasch G, Grünberg P, Saurenbach F, Zinn W. Enhanced magnetoresistance in layered magnetic structures with antiferromagnetic interlayer exchange. Phys Rev B 1989;39:4828−30.

[18] Baibich MN, Broto JM, Fert A, Nguyen Van Dau F, Petroff F, Etienne P, et al. Giant magnetoresistance of (001)Fe/(001)Cr magnetic superlattices. Phys Rev Lett 1988;61:2472−5.

[19] Parkin SSP, More N, Roche KP. Oscillations in exchange coupling and magnetoresistance in metallic superlattice structures: Co/Ru, Co/Cr, and Fe/Cr. Phys Rev Lett 1990;64:2304.

[20] Bakonyi I, Péter L. Electrodeposited multilayer films with giant magnetoresistance (GMR): progress and problems. Prog Mater Sci 2010;55(3):107−245.

[21] Belamkonda R, John T, Mathew B, DeCoster M, Hegab H, Davis D. Fabrication and testing of a CoNiCu/Cu CPP-GMR nanowire-based microfluidic biosensor. J Micromech Microeng 2010;20:25012−15.

[22] Wang SX, Li G. Advances in GMR biosensors with magnetic nanoparticle tags: review and outlook. IEEE Trans Magn 2008;44:1687−702.

[23] Sandhu A. Biosensing: new probes offer much faster results. Nat Nanotechnol 2007;2:746−8.

[24] Med Gadgets − internet journal of emerging medical technologies. 2008.

[25] Sawada S, Masuda Y, Zhu P, Koumoto K. Micropatterning of copper on a poly(ethylene terephthalate) substrate modified with a self-assembled monolayer. Langmuir 2006;22:332−7.

[26] Pogodina OA, Pustogov VV, de Melas F, Haberhauer-Troyer C, Rosenberg E, Puxbaum H, et al. Combination of sorption tube sampling and thermal desorption with hollow waveguide FT-IR spectroscopy for atmospheric trace gas analysis: determination of atmospheric ethane at the lower ppb level. Anal Chem 2004;76:464−8.

[27] Sabayev V, Croitoru N, Inberg A, Shacham-Diamand Y. The evolution and analysis of electrical percolation threshold in nanometer scale thin films deposited by electroless plating. Mater Chem Phys 2011;127:214−19.

[28] Shacham-Diamand Y, Inberg A, Sverdlov Y, Croitoru N. Electroless silver and silver-tungsten thin films, deposited for Microelectronics and micro-electro-mechanical system (MEMS) applications. J Electrochem Soc 2000;147:3345−9.

[29] Inberg A, Livshits P, Zalevsky Z, Shacham-Diamand Y. Electroless deposition of silver thin films on gold nanoparticles catalyst for micro and nanoelectronics applications. Microelectron Eng 2012;98:570−3.

[30] Livshits P, Inberg A, Shacham-Diamand Y, Malka D, Fleger Y, Zalevsky Z. Precipitation of gold nanoparticles on insulating surfaces for metallic ultra-thin film electroless deposition assistance. Appl Surf Sci 2012;258:7503−6.

[31] Mendlovic D, Zalevsky Z, Lohamnn AW. Various approaches in super resolution. Opt Photonics News 1997;8:21−2.

[32] Costa MFM. Application of image processing to the characterization of nanostructures. Rev Adv Mater Sci 2004;6:12−20.

[33] Duadi H, Livshits P, Gur E, Inberg A, Shacham-Diamand Y, Weiss A, et al. A novel algorithm to enhance blurred microscopy images of metallic ultra thin-films microstructures. J Microelectron Eng 2012;92:145−8.

[34] Johnson JL, Taylor JR. K-factor image factorization, Vol. 3715. SPIE; 1999; pp. 166−174

[35] Glickman E, Inberg A, Aviram G, Popovitz R, Croitoru N, Shacham-Diamand Y. Properties of 50 nm electroless films Ag–W–oxygen before and after low temperature, low activation energy resistivity decay. Microelectron Eng 2006;83:2359–63.

[36] Asher T, Inberg A, Glickman E, Fishelson N, Shacham-Diamand Y. Formation and characterization of low resistivity sub-100 nm copper films deposited by electroless on SAMT. Electrochim Acta 2009;54:6053–7.

[37] Allen G, Sorbello F, Altavilla C, Castorina A, Ciliberto E. Macro, micro and nano investigations on 3-aminopropyltrimethoxysilane self-assembly-monolayers. Thin Solid Films 2005;483:306–11.

CHAPTER *3*

New Super Resolving Techniques and Methods for Microelectronics

3.1 THE BASICS OF SUPER RESOLUTION

3.1.1 Introduction

Every imaging system presents a limited capability in resolution that can be expressed as a function of the minimal distance that two infinitely small spatial features can be positioned near each other while remaining two separable items in the image provided by the system [1].

New Approaches to Image Processing based Failure Analysis of Nano-Scale ULSI Devices.
DOI: http://dx.doi.org/10.1016/B978-0-323-24143-4.00003-3

But an imaging system is a medium that connects the input signal with the electronic output signal, leaving the detector. In that sense, the term "imaging system" must be divided into three parts, each one of which defines a different resolution limit.

First, we find the medium in which the optical signal propagates from the input plane through the optical system toward a detector. Here, the angular span of diffracted beams is linearly proportional to the optical wavelength and inversely proportional to the size of the feature that generates the diffraction. Thus, only the angles arriving within the diameter of the imaging lens are imaged at the detection plane, and the object's spectrum is trimmed by the limited aperture of the imaging system. The achievable resolution becomes restricted by diffraction and it is called *diffractive optical resolution* [2].

After that, the input signal is captured by a digital device (typically a coupled charge device [CCD] camera). Once again, the spatial information regarding the optical signal is distorted by detector array geometry. This limitation (named *geometrical resolution*) can be divided into two types of constraints. The first type is related to the number of sampling points and the distance between two such adjacent points. The denser the two-dimensional (2-D) spatial sampling grid is, the better the quality of the sampled image. The second limit is related to the spatial responsivity of each sampling pixel. Since each pixel integrates the light that impinges on its area, the 2-D sampling array is not an ideal sampling array. This local spatial integration done at each pixel results in a low pass filtered image [3].

After this process, the optical signal is converted into an electronic signal by digital conversion of the electrons collected in the electronic capacitor of every sampling pixel. In this third group, the quality of the detector comes into play and features such as sensitivity, dynamic range, and different types of noise will determine the quality of the final output signal [4]. The resolution defined in this stage is called *noise equivalent resolution*.

In any case, the resolution of the final electronic signal will be affected by the three previously defined factors. And resolution improvements of the final electronic readout signal can come from a combination of improvements achieved in each one of the three stages. The whole process of improvement is the real meaning of the term superresolution.

3.1.2 Fundamental limits to resolution improvement

Many researchers have used information theory to establish a relation between resolution and number of degrees of freedom of an optical system. Toraldo di Francia first proposed that the spatial frequency cut-off of an optical system could be extended [5]. Since then, a number of different schemes have been demonstrated to be true [6–19]. In 1966, Lukosz proposed an invariance theorem to explain the concepts underlying all superresolution approaches [10,11]. This theorem states that, for an optical system, it is not the spatial bandwidth, but the number of degrees of freedom (N_F) of the system that is fixed. That N_F value is, in fact, the number of points necessary to completely define the system in the absence of noise and is given by

$$N_F = 2(1 + L_x B_x)(1 + L_y B_y)(1 + T B_T), \qquad (3.1)$$

with B_x, B_y being the spatial bandwidths, L_x, L_y the dimensions of the field of view in the (x,y) directions, respectively, T the observation time, and B_T the temporal bandwidth of the optical system. Factor 2 is included because of the two independent, orthogonal polarization states.

Using this invariance theorem, that is, N_F = constant, Lukosz theorized that any parameter in the system could be extended above the classical limit if any other factor represented in Eq. 3.1 is proportionally reduced, provided that some *a priori* information concerning the object is known. However, this theorem is not complete because it does not consider noise in the optical system. Twenty years later, Cox and Sheppard included the noise factor in Lukosz's invariance theorem [15]. They considered the signal to noise ratio (SNR) as the noise factor and also considered three spatial dimensions: two independent polarization states and the temporal dimension.

$$N = (1 + 2L_x B_x)(1 + 2L_y B_y)(1 + 2L_z B_z)(1 + 2T B_T)\log(1 + SNR),$$

$$(3.2)$$

with being L_x, L_y, B_x, B_y, B_T, T defined in Eq. 3.1, L_z is the depth of field, and B_z is the spatial bandwidth in the z direction. Once again, factor 2 in each term of Eq. 3.2 refers to the two independent polarization states of each independent dimension.

Then, the invariance theorem of information theory states that it is not the spatial bandwidth, but the information capacity, of an imaging system that is constant. Thus, provided that the input object belongs to a restricted class, it is in principle possible to extend the spatial bandwidth (or any other desired parameter) by encoding-decoding additional spatial-frequency information onto the independent (unused) parameter(s) of the imaging system. And this is the fundamental principle underlying all superresolution schemes.

More recently, Mendlovic et al. introduced the space-bandwidth product adaptation (SW) concept as a tool for superresolution [20–24]. The SW concept generalizes the theorem of invariance of information capacity because not only degrees of freedom are considered, but also the shape of the SW of the imaging system. The SW can be applied to superresolution by considering that the SW of the optical signal is well adjusted with the SW of the imaging system. By adjusted we mean that the shape of the SW function of the signal is graphically contained within the shape of the SW function of the imaging system. Thus, the key point is to match the SW of the optical signal with the SW of the imaging system to allow effective transmission range of the optical signal through the imaging system.

But, as previously mentioned, the improvement of resolution requires *a priori* knowledge about the input object. One can use *a priori* information to classify the objects into different types allowing for different superresolution strategies. Thus, one finds angular multiplexing for non-extended objects [10,13], time multiplexing for temporally restricted objects [11,25,26], spectral encoding for wavelength restricted objects [8,27], spatial multiplexing with one-dimensional objects [11,12,28], polarization coding with polarization restricted objects [9,29,30], and gray level multiplexing for objects with restricted intensity dynamic range [31].

The previously presented outline deals with the strictly optical aspects of the imaging system (e.g., light propagation, lenses, object optical characteristics). A different aspect of the problem is the image recording by a sensor. The sensor, itself, introduces nonideal properties (mainly pixelation noise and sensitivity issues) that greatly influence the recorded image. In principle, a detector with large pixels is preferred, from the light collection capability point of view. Nevertheless, this results in a low-resolution image. This limitation is referenced as *geometrical resolution*. Several directions have been devised to overcome the geometrical resolution limit. They mostly deal with introducing subpixel information obtained by means of relative shift between the image and the sensor [32,33]. A small fill factor of the detector pixels can be used to further extend the superresolution capabilities [4,34].

3.1.3 Diffractive optical superresolution

The idea that resolving power is limited by diffraction in imaging systems, dates back to 1873 when German physicist Ernst Abbe published his theory that the resolution of an optical imaging system is limited by the wave nature of light [2]. Abbe reported on the role of wavelength and numerical aperture (NA) of lenses regarding microscope resolution in image formation. Since then, the ability to improve the resolving power of imaging systems beyond the limit imposed by diffraction has become a widely studied topic in applied optics.

Continuing Abbe's work, Helmholtz derived the maximum resolution achievable in microscopy, which can reach up to one-half of the illumination wavelength. Twenty years later, Lord Rayleigh published a manuscript with the mathematical foundation of Abbe's discovery. In 1906, Porter involved Fourier's theorem in the development of Abbe's theory and demonstrated the power of Fourier optics in image spatial filtering by placing different masks in the backfocal plane of a microscope lens [35]. Going back to the year 1874, Rayleigh reported on a quantitative way to define the resolution limit of an imaging system, and he established a resolution criterion which claims that two closed points having the same relative intensity are just resolved when the first minimum of the diffraction pattern in the image of one point source coincides with the maximum of the other. In such a case, the distance between the central maximum and the first minimum in the intensity point spread function (PSF) of the system is called a Rayleigh resolution distance. In paraxial approximation, the Rayleigh resolution distance is defined as $\delta X_{dif} = 1.22\lambda F/D$, when λ is the illumination wavelength, and F and D are the focal distance and the diameter of the imaging system. Equivalently, the Rayleigh resolution distance is $\delta X_{dif} \cong 0.6\lambda/NA$, when NA is the numerical aperture of the imaging system. Because the resolving power or resolution of an imaging system is defined as the inverse of the Rayleigh resolution distance, the previous definition provides us a direct way to increase the image resolution provided by an imaging system: by decreasing the illumination wavelength or by increasing the NA of the imaging lens, or both.

According to Abbe's theory, only a restricted cone of the spherical wave fronts emitted by a given point of the input object will be focused in its image by an optical imaging system. In a general sense, only a portion of the diffracted components generated when an object is illuminated can pass through an optical system due to the limited size of the input pupil. Figure 3.1 depicts this situation.

From a spatial-frequency domain point of view, a wide analysis of diffraction limited optical systems allows the definition of a transfer function representative of the optical system transmittance in the Fourier domain [1]. Such frequency response implies the definition of a cut-off frequency that means a truncation in the spatial-frequency content of the input object. The cut-off frequency is defined (for coherent and incoherent illumination) in terms of the NA of the imaging system and the illumination wavelength λ as

$$f_c^{coh} = \frac{NA}{n\lambda} \quad and \quad f_c^{incoh} = \frac{2NA}{n\lambda}, \quad (3.3)$$

With n being the refractive index of the medium between the object and the imaging system. Once again, in order to make expand the transfer function, or to improve the spatial cut-off frequency, we must increase the NA of the optical system or decrease the illumination wavelength or both. But this procedure is not always possible, and it is necessary to define new techniques that will be able to improve the resolution without changes in the physical properties of the optical system, that is, without modifications in the following parameters: diameter, focal length, and illumination wavelength. These new superresolution techniques can be understood to be a definition of a synthetic numerical aperture (SNA) higher than the conventional NA, which implies a reduction in the

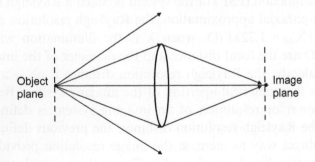

Figure 3.1 Image formation in the scope of the Abbe's theory: the object's spectrum is trimmed by the system's aperture when on-axis illumination is used.

Rayleigh resolution distance and, therefore, an improvement in the resolution of the imaging system. Figure 3.2 is representative of the SNA.

The key to generating an SNA is the introduction of additional optical elements (or masks) or the use of a special illumination procedure in the imaging system to help us to achieve the superresolution effect. Some examples of such masks are diffraction gratings (physical or projected) or prisms, while tilted beams are the most utilized illumination procedure. The selection of the aided optical element is related to a priori knowledge about the input object, that is, to the invariance theorem of information theory. In any case, either the masks used in the encoding should have a critical size below the diffraction limit of the imaging system or the illumination angle of the tilted beam must be higher than the imaging lens NA. Then, it is possible to divert the object's high spatial frequency content toward the system's limited aperture, allowing its transmission through it. After that, any additional information must be recovered and placed, via the proper decoding procedure, to its original position into the object's spectrum.

In general, the superresolving effect is achieved in at least two basic steps. First, an encoding process over the object's spectrum is needed in the sense that higher spatial-frequency content can go through the imaging lens aperture. The encoding is performed by the mask that has been selected for it. Let us call this stage the *encoding stage*. After that, a second stage, named the *decoding stage*, is needed to allow recovery and correct replacement of the additional frequency content transmitted due to the encoding stage. Additionally in some cases, the relocation of the new band-pass frequency content to its original position in the object's spectrum cannot be performed during the decoding

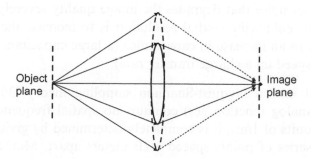

Figure 3.2 Generation of the SNA (white dashed big lens) in comparison with the conventional aperture (solid small lens).

procedure. Then, a third stage, named the *digital post-processing stage*, is needed to allow final superresolved imaging. But because digital sensors are relatively modern developments, historically superresolution approaches were only composed of the two first stages.

3.1.4 Geometrical superresolution

One of the basic resolution limitations is related to the geometry of the sensing array. This limitation is affected both by the number and the pitch of the sampling pixels (i.e., the density of the sampling grid, as well as the spatial responsivity of each pixel). The spatial responsivity determines the point spread function, while the pitch is related to the Nyquist sampling condition. Thus, the ideal sampling case is to be obtained through a large number of pixels with a small pitch, while each pixel is a delta function. That way—due to the small pitch—the Nyquist frequency can be high enough to recover all the spectral content in the image. And if the sampling pixels are delta functions, this is an ideal sampling with no low passing effects generated due to the spatial responsivity of the sampling points. However, due to practical reasons of energetic efficiency, this is not the geometrical structure of common available sensors. Thus, when one refers to geometrical superresolution he or she refers to approaches aiming to overcome those two factors.

3.1.4.1 Sampling density

As previously mentioned, the most direct solution for increasing spatial resolution is to reduce the pixel size (i.e., increase the number of pixels per unit area) by sensor manufacturing techniques. As the pixel size decreases, however, the amount of light available also decreases. This generates shot noise that degrades the image quality severely. Another approach for enhancing spatial resolution is to increase the chip size, which leads to an increase in capacitance (a large capacitance makes it difficult to speed up a charge transfer rate).

The well known Nyquist-Shannon sampling theory [36,37] states that if an analog function $g(x)$ contains no spatial frequencies higher than B (at units of $1/m$), it is completely determined by giving its ordinates at a series of points spaced $1/2B$ meters apart. Mathematically, spatially sampling the signal $g(x)$ means its multiplication with a train of Dirac's delta functions:

$$g_s(x) = g(x) \sum_n \delta(x - nx_s),$$ (3.4)

where x_s is the spatial sampling pitch (the distance between the sampling points) and δ is the delta of Dirac.

The spatial spectrum (i.e., the Fourier transform of the sampled signal $g(x)$ or $g(nx_s)$), equals:

$$G_s(\mu) = \int g_s(x)\exp(-2\pi ix\mu)dx = \sum_n G\left(\mu - \frac{n}{x_s}\right),$$ (3.5)

where $G(\mu)$ is the spatial spectrum of the original signal $g(x)$ and μ is the spectral coordinate. If the original signal $g(x)$ is band limited by having no spatial frequencies higher than B, this means that the replications become nonoverlapping when:

$$\frac{1}{x_s} \geq 2B.$$ (3.6)

In this case, the reconstruction of the original signal $g(x)$ from its sampled version is obtained by filtering only one replica in the Fourier domain and then performing the inverse Fourier transform in order to come back to the space domain.

In case the geometrical resolution of the sampled image is reduced due to insufficiently dense spatial sampling (i.e. the condition of Eq. 3.6 is not fulfilled), the basic solution is to capture a set of signals while each is performed at a given reduced sampling and with proper subpixel shift, to properly interlace the samples and then to construct new signals containing sampling corresponding to the required higher sampling rate. In optics, this procedure is called microscanning since the subpixel shifts are obtained by mechanical scanning of mirrors reflecting the incoming radiation toward the detection array [38]. Figure 3.3 presents the described technique when the sampling rate is improved by a factor of 4.

3.1.4.2 Nonideal sampling
The second effect that is directly related to the reduction of geometrical resolution is connected to the function used to perform the sampling, rather than to the density of the sampling itself. As previously stated, the ideal sampling is based upon a train of Dirac's delta functions. However, in practice, the sampling is usually done with detectors having areas that are far from being delta functions. Each pixel of the

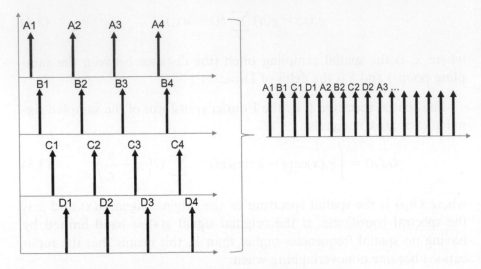

Fig. 3.3 Sampling with subpixel shifting. Demonstration for improvement factor of 4.

sampling detector averages the radiation that arrives to its area. This spatial averaging operation reduces the geometrical resolution of the imaging sensing array. If we denote the size of each sampling pixel by δx, then the signal $g(x)$—after being sampled by such a nonideal sampling—can be described as:

$$g_s^{(p)}(x) = \left[\int g(x')rect\left(\frac{x-x'}{\delta x}\right)dx \right]\left[\sum_n \delta(x - nx_s) \right], \qquad (3.7)$$

where we assumed that each pixel collects light within a rectangle with a size of δx. Obviously, if the responsivity of the pixels is not uniform, instead of a rectangle, we will use the spatial distribution of the pixels' responsivities. $g_s^{(p)}$ denotes the signal $g(x)$ being sampled with a nonideal sampling function. If we perform a Fourier transform over the expression of Eq. 3.7, we obtain (up to a proportion constant) the following expression:

$$G_s^{(p)}(\mu) = [G(\mu)\text{sinc}(\delta x\mu)] \otimes \left[\sum_n \delta\left(\mu - \frac{n}{x_s}\right) \right]$$

$$= G\left(\mu - \frac{n}{x_s}\right)\text{sinc}\left(\delta x\mu - n\frac{\delta x}{x_s}\right), \qquad (3.8)$$

where $\otimes$ denotes the convolution operation and *sinc* is the Fourier transform of a *rect* function that is defined as:

$$\text{sinc}(\delta x \mu) = \frac{\sin(\pi \delta x \mu)}{\pi \delta x \mu}. \tag{3.9}$$

Therefore, the meaning of the nonideal sampling, as seen from Eq. 3.8, is that the spectrum of g(x) is low passed with the *sinc* function and thus the bandwidth of the information is now limited by $1/\delta x$ and not by the real bandwidth of g(x). Thus, the geometrical superresolution techniques dealing with this limitation aim to overcome the bandwidth reduction imposed by the *sinc*. There are several types of approaches that aim to overcome this type of geometrical resolution limitation.

One possible way to consider the problem of resolution enhancement is by looking at each pixel in the sampling array as a subpixel matrix functioning as an averaging operator, so that for each shift of the camera, some new subpixels are added and some are removed from the averaging. This is equivalent to the convolution of a superresolved image and a function having the spatial shape of a pixel. In this way, it is possible to solve the problem of image enhancement by inverse filtering that is realized by dividing the frequency domain by the Fourier transform of the pixel's spatial responsivity distribution (i.e. pixel's spatial shape). The solution to this problem of inverse filtering is not possible when there are zeros in this inverse function (one over the Fourier of the pixel's shape). The addition of a special periodic spatial mask—attached to the detection array or positioned in the intermediate image plane (the period is the pitch of the pixels and thus the same transmission is added to each one of the pixels)—can modify the Fourier of the pixel's shape and allow the realization of the inverse filtering operation [34]. Another possible solution is to use a pseudorandom phased mask attached to the imaging lens [39] (in optical based imaging).

3.1.4.3 Image sequence approaches
The limitation which is related to the spatial density of the pixels can easily be removed by a procedure called microscanning and interlacing (i.e., adding many low resolution images, each taken at slightly different small geometrical shifts) [38]. Originally, this increased the resolution for 1-D images. Similarly, it is possible to interlace images in 2-D and to obtain higher resolution. The resolution improvement can be

obtained, not only by generating a controlled movement between the camera and the object, but also by using the existing jittering of the camera and applying registration algorithms [40].

This solution of interlacing and camera shifting is not complete due to the frequency folding phenomenon (spectral aliasing) that blurs high frequencies. Much effort has been put in in the last few years to remedy this problem [41].

Note that another relevant aspect of improving the geometrical resolution is related to the blurring generated in the image due to the movement of the target while performing the time integration of the sampling sensor. This deblurring may be recovered by using a sequence of images [42,43].

An important set of digital processing approaches providing superresolution images by solving the inverse problem while using a sequence of images and estimating the existing noises and the blurring function, was developed by Elad et al. [44−46].

3.1.4.4 Approach involving physical components

The limitation related to the shape of each sampling pixel can be resolved by applying various approaches. One way of considering the problem of resolution enhancement is by looking at each pixel in the sampling array as a subpixel matrix functioning as an averaging operator, so that for each shift of the camera, some new subpixels are added to and some are removed from the averaging. This is equivalent to the convolution of a superresolution image and a function having the spatial shape of a pixel. In this way, it is possible to solve the problem of image enhancement by inverse filtering: dividing in the frequency domain using the Fourier transform of the pixel's shape such that no more zeros will exist anymore in the Fourier domain and this will allow doing the inverse filtering operation [34,47]. Another possible solution is to use a pseudorandom phased mask attached to the lens [39].

Instead of physically attaching the mask to the sensor, one may use projection to create the desired distribution on top of the object [48].

Another physical-related technique involves the positioning of the physical mask—not near the sensor—but rather in the intermediate image plane. Due to the spatial blurring generated by the spatial responsivity of the pixels of the sensor, one actually has more variables to recover (the high resolution information) than the number of equations (the sampled readout coming from the detector). Thus the insertion of the mask inserts *a priori* knowledge that increases the number of equations, allowing us eventually to perform improved reconstruction of the high resolution image out of its blurred version by applying the matrix inversion procedure [49].

3.1.4.5 Digital processing methods

A related problem to superresolution (SR) techniques is image restoration, which is a well-established area in image processing applications [50,51]. The goal of image restoration is to recover a degraded (e.g., blurred, noisy) image, not to change the size of the image. Another problem related to SR reconstruction is image interpolation that has been used to increase the size of a single image [52−54].

The common observation model involves resolution degradation due to a warp matrix (the motion that occurs during image acquisition), a blur matrix (caused by an optical system being out of focus or by a diffraction limit or by optical aberrations, or by relative motion existing between the imaging system and the original scene) and a subsampling matrix (which generates aliased low resolution (LR) images from the warped and blurred high resolution (HR) image)—all in the presence of additive noise.

The reconstruction process involves: registration (the relative shifts between LR images compared to the reference LR image are estimated with fractional pixel accuracy), interpolation (nonuniform interpolation is necessary to obtain a uniformly spaced HR image from a nonuniformly spaced composite of LR images), and restoration (i.e., inverse procedure/deconvolution). The estimation of motion information is referred to as registration [55−58]. The restoration is applied to the upsampled image to remove blurring and noise.

Ur and Gross [59] performed a nonuniform interpolation of an ensemble of spatially shifted LR images by utilizing the generalized multichannel sampling theorem of Papoulis [60] and Brown [61]. Komatsu et al. [62] presented a scheme to acquire an improved resolution image by applying the Landweber [63] algorithm from multiple images taken simultaneously with multiple cameras with different apertures [64]. Hardie and colleagues developed a technique for real-time infrared image registration and SR reconstruction [65] doing a gradient-based registration algorithm for estimating the shifts between the acquired frames and presented a weighted nearest neighbor interpolation approach. Finally, Wiener filtering is applied to reduce the effects of blurring and noise caused by the system.

The frequency domain approach makes explicit use of the aliasing that exists in each LR image to reconstruct an HR image [66]. The frequency domain approach is based on the following three principles: i) the shifting property of the Fourier transform (shifting in space results in a linear phase factor in the Fourier), ii) the aliasing relationship between the continuous Fourier transform (CFT) of an original HR image and the discrete Fourier transform (DFT) of observed LR images, and iii) the assumption that an original HR image is band limited.

$$Y = \Phi X, \tag{3.10}$$

where Y is a column vector with the DFT coefficients, X is a column vector with the samples of the unknown CFT, and Φ is a matrix that relates the DFT of the observed LR images to samples of the continuous HR image. Therefore, the reconstruction of a desired HR image requires us to determine Φ and solve this inverse problem. The Tikhonov regularization method is adopted to overcome the ill-posed problem resulting from the blur operator.

Bose et al. [67] proposed the recursive total least squares method for SR reconstruction to reduce effects of registration errors (errors in Φ). A discrete cosine transform (DCT)-based method was proposed by Rhee and Kang [68] where the processing memory requirements were reduced and multichannel adaptive regularization parameters were applied to overcome ill-posedness, such as in underdetermined cases or in insufficient motion information cases. In the case of regularization, there are deterministic and stochastic approaches. Typically, constrained least squares (CLS) and maximum *a posteriori* (MAP) SR image reconstruction methods are introduced. In deterministic approaches, the iterative solution is commonly used in which:

$$\hat{x}^{(n+1)} = \hat{x}^{(n)} + \beta \left[\sum_{k=1}^{p} W_k^T (y_k - W_k \hat{x}^{(n)}) - \alpha C^T C \hat{x}^{(n)} \right], \tag{3.11}$$

where the operator C is generally a high-pass filter; W represents—via blurring, motion, and subsampling—the contribution of HR pixels in x to the LR pixels in y_k ($k = 1...p$); α represents the Lagrange multiplier, commonly referred to as the regularization parameter, which controls the tradeoff between fidelity to the data and smoothness of the solution; and β represents the convergence parameter. W_K^T contains an upsampling operator and a type of blur and warping operator. Katsaggelos and colleagues [69,70] proposed a multichannel regularized SR approach in which the regularization functional is used to calculate the regularization parameter without any prior knowledge at each iteration step. Later, Kang formulated the generalized multichannel deconvolution method, including the multichannel regularized SR approach [71]. The SR reconstruction method obtained by minimizing a regularized cost functional was proposed by Hardie et al. [72].

Stochastic SR image reconstruction, typically a Bayesian approach, provides a flexible and convenient way to model *a priori* knowledge concerning the solution. Bayesian estimation methods are used when the *a posteriori* probability density function (PDF) of the original image can be established. The MAP estimator of x maximizes the *a posteriori* PDF $P(x|y_k)$ with respect to x:

$$x = \arg \max P(x|y_1, y_2..., y_p) = \arg \max\{\ln P(y_1, y_2..., y_p|x) + \ln P(x)\}.$$

$$(3.12)$$

A maximum likelihood (ML) estimation has also been applied to the SR reconstruction. The ML estimation is a special case of MAP estimation with no prior term. Due to the ill-posed nature of SR inverse problems, however, MAP estimation is usually used in preference to ML. Tom and Katsaggelos [73] proposed the ML SR image estimation problem to estimate the subpixel shifts, the noise variances of each image, and the HR image simultaneously. The SR reconstruction from an LR video sequence using the MAP technique was proposed by Schultz and Stevenson [74]. Robustness and flexibility in modeling noise characteristics and *a priori* knowledge about the solution are the major advantages of the stochastic SR approach.

The Projection onto Convex Sets (POCS) method describes an alternative iterative approach to incorporate prior knowledge about the solution into the reconstruction process. With the estimation of registration parameters, this algorithm simultaneously solves the restoration and interpolation problems to obtain the SR image. The POCS formulation of the SR reconstruction was first suggested by Stark and Oskoui [75]. The method of POCS can be applied to find a vector which belongs in the intersection by the recursion:

$$x^{(n+1)} = P_m P_{m-1} \ldots P_2 P_1 \, x^{(n)}, \tag{3.13}$$

where x_0 is an arbitrary starting point, and P_i ($i = 1, 2, \ldots, m$) is the projection operator which projects an arbitrary signal x onto the closed, convex sets, C_i ($i = 1,2,\ldots,m$) that are defined as a set of vectors that satisfy a particular property.

Other SR approaches include the Iterative Back-Projection Approach. Irani and Peleg [76] formulated the iterative back-projection (IBP) SR reconstruction approach that is similar to the back projection used in tomography. In this approach, the HR image is estimated by back projecting the error (difference) between simulated LR images via the imaging blur and the observed LR images. This process is repeated iteratively to minimize the energy of the error. Elad and Feuer [77] proposed an SR image reconstruction algorithm based on adaptive filtering theory applied in a time axis. They modified the notation in the observation model to accommodate for its dependence on time and suggested least squares (LS) estimators. The SR reconstruction algorithms presented so far require relative subpixel motions between the observed images. However, it is shown that SR reconstruction is also possible from differently blurred images without relative motion [44,78]. Rajan and Chaudhuri [79] presented the SR method using photometric cues, and the SR technique using zoom as a cue is proposed by Joshi and Chaudhuri [80].

In most SR reconstruction algorithms, the blurring process is assumed to be known. In many practical situations, however, the blurring process is generally unknown or is known only to within a set of parameters. Therefore, it is necessary to incorporate the blur identification into the reconstruction procedure. Wirawan et al. [81] proposed a blind multichannel SR algorithm by using multiple finite impulse response (FIR) filters.

The blurring problem is part of the inverse problem. However, in order to be defined as well-posed and to have a unique solution, it must uphold the following three conditions: existence, stability, and uniqueness. If some of the conditions do not hold, then the problem is ill-posed, and there may not be a solution, or it may not be unique

[82]. Furthermore since this solution does not uphold all three conditions mentioned above, the additive noise prevents us from converging to a real solution. Likewise, since there are more unknowns than equations, the solution is not unique. Finally, a small change in one of the variables would affect the solution of the problem so that the stability of the solution would be very low.

One possible direction for the above mentioned problems is to use the pseudo inverse matrix that is obtained by reduction of the least square error. Techniques dealing with least square error reduction [83] involve recursive least square error (RLS) [84] and recursive total least square error (RTLS) [85]. A more sophisticated method to reduce least square errors recursively uses regularization [86]. This method, which succeeds in overcoming noise, contains Tikhonov's regularization component. This component is designed such that for problems without noise, it will be possible to reduce this component so the real solution will be approached, while for images with noise, this positive addition will yield an optimal solution [87].

A further way of describing the convolution between the superresolution image and the spatial shape of the pixel, is by a set of linear equations $A\hat{x} \approx b$, where $\hat{x}$ represents the super-resolution image translated to a vector, b represents the interlaced discrete image constructed as a spread vector, and A represents a matrix spreading its convolution action caused due to the spatially extended shape of the pixels. Such a mathematical description is feasible since the geometrical shape of the pixels can be represented as a convolution operation that blurs the spatial resolution of the original image. This is due to the local averaging performed within the area of each pixel. Apparently, also here, relating to the above problem as a set of equations, the simple answer to our problem is to inverse the convolution matrix and to extract the estimator $\hat{x}$ as $\hat{x} = bA^{-1}$. Since there is no existing solution, one may use the method of reducing the least square error $\|Ax - b\|^2$, such that we shall obtain a pseudo inverse:

$$\hat{x} = (A^T A)^{-1} A^T b, \tag{3.14}$$

or using Tikhonov's regularization [87,88] and obtaining:

$$\hat{x} = (A^T A + \alpha^2 I)^{-1} A^T b \tag{3.15}$$

3.1.5 Wigner Transform

In this subsection, we describe a mathematical tool called the Wigner transform [24,89,90] and use it to represent the two geometric resolution limits mentioned in the previous section. Then, we use Wigner to represent the geometric super resolving approaches that involve applying random masks for encoding, as well as microscanning for increasing the spatial sampling density. The Wigner transform is a phase space representation, which is basically a bilinear spectrogram of the signal. In our case the signal is a spatial signal, and the Wigner distribution is a spectrogram having two axes of the space and the spatial frequency.

3.1.5.1 Wigner of sampled signals

The Wigner distribution of a sampled signal can be computed as follows. We will denote the sampled version of $f(x)$ by $f_s(x)$:

$$f_s(x) = f(x) \cdot \sum_{n=-\infty}^{\infty} \delta(x - n\delta x) = \sum_{n=-\infty}^{\infty} f(n\delta x)\delta(x - n\delta x), \qquad (3.16)$$

where δx is the spatial sampling interval and $\delta(x)$ is the delta function of Dirac. The Wigner distribution of the sampled signal of Eq. 3.16 is:

$$W_{f_s}(x, \nu) = \int f_s\left(x + \frac{x'}{2}\right)f_s^*\left(x - \frac{x'}{2}\right)\exp(-2\pi i\nu x')dx'$$

$$= \sum_n \sum_m f(n\delta x)f^*(m\delta x)\delta\left(x - \left(\frac{n+m}{2}\right)\delta x\right) \qquad (3.17)$$

$$\exp(-2\pi i\nu(n-m)\delta x)$$

Let us assume for our numerical simulation a Gaussian signal of:

$$f(x) = \frac{1}{\sqrt{2\pi}\sigma}\exp\left(-(x-\Delta x)^2/2\sigma^2\right). \qquad (3.18)$$

where σ is the standard deviation of the Gaussian and Δx is its average. We will assume in our Matlab simulation a signal with 1024 pixels while Δx equals 512 pixels. The standard deviation we chose was 32 pixels. The result of the Wigner simulation is seen in Fig. 3.4(a). When performing under sampling, the resultant Wigner distribution can be seen in Fig. 3.4(b). There the sampling period δx was of 32 pixels. One may see that the spectral distribution of Fig. 3.4(b) is indeed extended in comparison to that of Fig. 3.4(a) due to the (overlapping) spectral

(a)

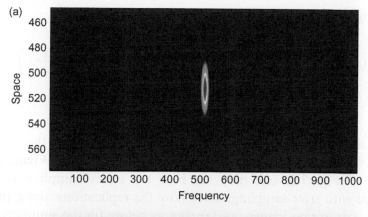

(b)

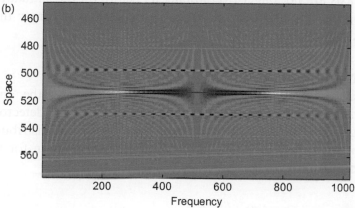

Figure 3.4 (a) Wigner transform of a Gaussian signal. (b) Wigner transform of under sampled signal.

replications. Although over the spectral axis the distribution is very wide, the high resolution information is lost since the different spectral replications overlap and thus superimpose/mix.

3.1.5.2 Wigner of microscanned signals

As previously explained, microscanning by a factor of M can resolve the under sampling problem since it increases the density of the spatial sampling. Denoting the microscanned signal as $g_M(x)$ yields:

$$
\begin{aligned}
g_M(x) &= \sum_{m=0}^{M-1}\left[\sum_{n=-\infty}^{\infty} f\left(n\delta x - m\frac{\delta x}{M}\right)\delta\left(x - n\delta x - m\frac{\delta x}{M}\right)\right]\\
&= \sum_{n=-\infty}^{\infty} f\left(n\frac{\delta x}{M}\right)\delta\left(x - n\frac{\delta x}{M}\right).
\end{aligned}
\tag{3.19}
$$

The Wigner distribution of the microscanned signal is:

$$W_{g_M}(x,\nu) = \sum_n \sum_m f\left(\frac{n}{M}\delta x\right) f^*\left(\frac{m}{M}\delta x\right) \delta\left(x - \left(\frac{n+m}{2M}\right)\delta x\right)$$
$$\exp\left(-2\pi i\nu(n-m)\frac{\delta x}{M}\right).$$

(3.20)

In the simulation of Fig. 3.5, we used M = 16, such that the sampling period was effectively reduced to $\delta x/M = 2$ pixels. In the Wigner distribution of Fig. 3.5, one may see how the under sampling is being replaced with over sampling, since now the replications along the frequency axis are no longer overlapping as before (in the simulated over sampling the distance between spectral replications is larger than the bandwidth of the signal).

3.1.5.3 Nonideal sampling

Nonideal sampling is basically equivalent to blurring the analogue function with the spatial responsivity of the pixels of the detector prior to being ideally sampled. Denoting by r(x) the spatial distribution of the responsivity of the pixels, yields the following relation for the not ideally sampled signal f(x):

$$f^{(b)}(x) = \int f(x')r(x - x')dx'.$$

(3.21)

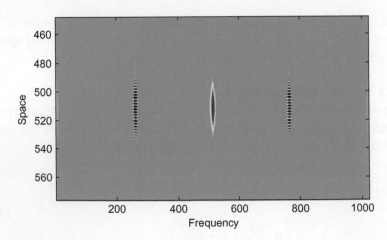

Figure 3.5 Wigner transform of a sampled signal after performing microscanning.

After the blurring, the signal is ideally sampled, which yields:

$$f^{(b)}(x) = \sum_{n=-\infty}^{\infty} f^{(b)}(n\delta x)\delta(x - n\delta x) = \sum_{n=-\infty}^{\infty} \int f(x')r(n\delta x - x')\delta(x - n\delta x)dx'.$$

$$(3.22)$$

The Fourier transform of the not ideally sampled signal equals:

$$F_{fs}{}^{(b)}(\nu) = (F(\nu)R(\nu)) \otimes \frac{1}{\delta x}\sum_{n}\delta\left(\nu - \frac{n}{\delta x}\right)$$

$$= \frac{1}{\delta x}\sum_{n}\int F(\nu')R(\nu')\delta\left(\nu - \nu' + \frac{n}{\delta x}\right)d\nu', \qquad (3.23)$$

$$= \frac{1}{\delta x}\sum_{n}F\left(\nu + \frac{n}{\delta x}\right)R\left(\nu + \frac{n}{\delta x}\right)$$

where $\otimes$ denotes a convolution operation and the capital letters $F(\nu)$ and $R(\nu)$ stand for the Fourier transform of the spatial signals $f(x)$ and $r(x)$:

$$F(\nu) = \int f(x)\exp(-2\pi ix\nu)dx \quad R(\nu) = \int r(x)\exp(-2\pi ix\nu)dx \quad (3.24)$$

Thus, the nonideal sampling yields low pass filtering in the Fourier domain, being imposed by $R(\nu)$. Therefore, the improvement of anti aliasing obtained before due to the microscanning is now being lost due to this low pass filter. The Wigner transform of the nonideally sampled signal is:

$$W_{fs^{(b)}}(x, \nu) = \int F_{fs}^{(b)}\left(\nu + \frac{\nu'}{2}\right)F_{fs}^{(b)*}\left(\nu - \frac{\nu'}{2}\right)\exp(2\pi i\nu'x)d\nu'$$

$$= \frac{1}{\delta x^2}\int \sum_{n=-\infty}^{\infty}\sum_{m=-\infty}^{\infty}F\left(\nu + \frac{\nu'}{2} + \frac{n}{\delta x}\right)R\left(\nu + \frac{\nu'}{2} + \frac{n}{\delta x}\right)$$

$$F^*\left(\nu - \frac{\nu'}{2} + \frac{m}{\delta x}\right)R^*\left(\nu - \frac{\nu'}{2} + \frac{m}{\delta x}\right)\exp(2\pi i\nu'x)d\nu'$$

$$(3.25)$$

Note that in this case we used the definition of the Wigner based upon the Fourier transform of the signals (unlike the definition of Eq. 3.17, where the spatial distribution of the signal was used to compute the Wigner transformation).

In Fig. 3.6, we present the numerical simulation of the nonideal sampled signal f(x). One may see how the low pass effect of the non-ideal sampling reduces the high frequency content (the replicated spectral information is much narrower along the frequency axis). In the simulation of Fig. 3.6, we blurred the spatial Gaussian f(x) with spatial *rect* function r(x) having the width of 128 pixels and that represents detection pixels having the dimension of 128 pixels each. The performed microscanning was such that $\delta x/M = 2$ pixels. One may see in Fig. 3.6 that, indeed, due to the microscanning, the spectral replications are sufficiently separated, but due to the nonideal sampling, each replication is low passed, and thus it contains less spectral frequencies than the original signal.

3.1.5.4 Geometrical super resolution
As previously mentioned, the geometric superresolution configuration that overcomes the nonideal sampling of the pixels of the detection array, involves the addition of a spatially random encoding mask positioned in the intermediate image plane. This encodes the analogue light distribution prior to its being sampled and microscanned (to increase the obtainable spatial sampling density). Usage of spatial random encoding prior to nonideal sampling can recover the loss we obtained in Fig. 3.6 for the high frequencies. We assume that our high resolution spatial encoding mask is denoted by p(x) yielding the following nonideal sampled signal (after being encoded):

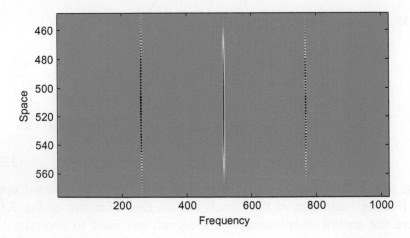

Figure 3.6 Wigner transform of nonideal sampled signal after performing micro scanning.

$$f_s^{(b,e)}(x) = \left[\int (f(x')p(x'))r(x-x')dx' \right] \sum_{n=-\infty}^{\infty} \delta(x-n\delta x), \qquad (3.26)$$

and its Fourier transform equals:

$$F_{fs}^{(b,e)}(\nu) = \left[\int (F(\nu')P(\nu-\nu')d\nu')R(\nu) \right] \otimes \frac{1}{\delta x} \sum_{n=-\infty}^{\infty} \delta\left(\nu - \frac{n}{\delta x}\right)$$

$$= [(F(\nu') \otimes P(\nu-\nu'))R(\nu)] \otimes \frac{1}{\delta x} \sum_{n=-\infty}^{\infty} \delta\left(\nu - \frac{n}{\delta x}\right) \qquad (3.27)$$

where $\otimes$ denotes a convolution operation. This results in:

$$F_{fs}^{(b,e)}(\nu) = \frac{1}{\delta x} \sum_{n=-\infty}^{\infty} \int\int F(\nu'')P(\nu'-\nu'')R(\nu')\delta\left(\nu-\nu'+\frac{n}{\delta x}\right)d\nu'd\nu''$$

$$= \frac{1}{\delta x} \sum_{n} R\left(\nu + \frac{n}{\delta x}\right) \int F(\nu'')P\left(\nu + \frac{n}{\delta x} - \nu''\right)d\nu''$$

$$(3.28)$$

By applying the Wigner definition based upon the Fourier transform of signals (Eq. 3.25), we may compute the resulting Wigner distribution of $f_s^{(b,e)}(x)$:

$$W_{f_s^{(b,e)}}(x,\nu) = \frac{1}{\delta x^2} \sum_{n} \sum_{m} R\left(\nu + \frac{\nu'}{2} + \frac{n}{\delta x}\right) R^*\left(\nu - \frac{\nu'}{2} + \frac{m}{\delta x}\right)$$

$$\iiint F(\nu_1)P\left(\nu + \frac{\nu'}{2} + \frac{n}{\delta x} - \nu_1\right)$$

$$\cdot F^*(\nu_2)P^*\left(\nu - \frac{\nu'}{2} + \frac{m}{\delta x} - \nu_2\right) \exp(2\pi i\nu' x)d\nu_1 d\nu_2 d\nu'$$

$$(3.29)$$

Since now the spatial signal is encoded before being low passed, the information is not lost, but rather spread over the entire spectral domain. Right afterward, each replication (which contains the full spectral information due to the encoding) after solving the aliasing problem via the microscanning may be used to decode the resolution information of the original high resolution signal f(x). For the simulation of Fig. 3.7(a), we used a binary random high resolution encoding mask having the Wigner transform seen in Fig. 3.7(b). The microscanning operation generated the sampling period of $\delta x/M = 2$ pixels. One

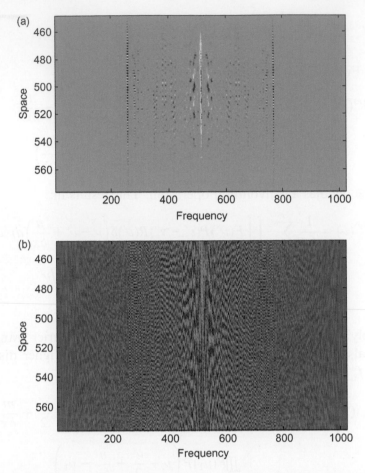

Figure 3.7 (a) Wigner transform of encoded signal after performing nonideal sampling. (b) Wigner transform of the spatial binary random encoding mask.

may see how the information is spread over the entire Wigner domain due to the high resolution binary encoding mask.

Since the Wigner definition may be computed either using spatial signals (Eq. 3.17) or the Fourier transform of signals (see Eq. 3.25), performing the encoding over the Fourier (spectral) domain rather than over the space may result in a similar outcome of spreading, rather than losing the high resolution information. In spectral encoding, one needs to perform decoding as well, in order to recover the mixed information. From Fig. 3.7, one may see that, unlike in Fig. 3.6, the information is spread over the entire phase space; however, its spreading is not a simple replication, but rather expresses the encoding that was applied. This encoding (expressed as information spreading in the Wigner space) will enable us perform to the decoding and to obtain the information reconstruction.

3.2 SUPER-RESOLUTION IMAGING FOR IMPROVED FAILURE ANALYSIS

With the advent of multimetal layer processes, flip chip packaging, and the continuous shrinkage of critical geometries, the ability to access internal nodes during probing and silicon microsurgery has become increasingly difficult. The resolution limit is an order of magnitude above the device feature densities in the < 90 nm era when using air gap optics, and it improves by a factor of 4 when using solid immersion optics. The scaling down of transistor geometry is leading to the inevitable consequence where more than 50% of the transistors in a 45 nm process have widths smaller than 0.15 um.

The first limitation that is encountered by the analyst is the difficulty in resolving small geometries from the backside. The active regions look like gray clouds already blurring into 0.13 um designs. Nevertheless, the use of accurate stages and CAD (computer aided design) overlay capability, which are essentially incorporated into any modern backside tool, could alleviate spatial resolution issues and allow blind navigation.

In their 1972 paper, Gerchberg and Saxton [91] demonstrated an iterative algorithm for phase retrieval. After performing a Fourier transform of an image, one might have access only to the magnitudes (or intensities) of the object and the Fourier image and not to the phase distribution. Since the phase distribution is vital to reproducing the Fourier image (or to performing a correct inverse Fourier transform), it is important to know how to retrieve the lost data, as phase in the object plane translates also to magnitude in the Fourier plane and vice versa.

In their work, Gerchberg and Saxton showed that if one knows the complete magnitude distribution in both the object plane and the Fourier plane, then one can extract the phase distribution in both planes. This is done by imposing the magnitude at the object plane, performing a Fourier transform, imposing the magnitude at the Fourier plane (keeping the phase distribution), performing an inverse Fourier transform, imposing the magnitude at the object plane (keeping the phase distribution), and so on. Although this iterative procedure does not ensure that the result obtained is optimal, it does suggest a solution, and the procedure always converges. Gerchberg [92] and later Papoulis [93] used the concept behind the original Gerchberg–Saxton algorithm to reconstruct an object when a certain region of the distribution is known in the object plane and a certain region is known in the Fourier plane. The Gerchberg–Papoulis algorithm is quite similar to the original Gerchberg–Saxton algorithm, except for imposing field distributions in specific regions instead of magnitude-only distributions in the entire region of discourse.

In this subsection, we will use an iterative algorithm based upon the numerical approach of Gerchberg–Papoulis in order to improve the resolution of microelectronic chip imaging and to perform its improved failure analysis.

3.2.1 Resolution limit in failure analysis

Shrinking device geometries in integrated circuits are made possible by improving optical lithography, which is driven mainly by decreasing the wavelength of the exposure source. However, optical probing relies on the transmission of light through the silicon substrate. Therefore, due to the strong absorption at photon energies greater than the band gap of Si, the wavelength cannot be less than the band gap (~ 1.1 um). As a result, backside tools have inherent degraded spatial resolution.

The spatial resolution of a high-resolution microscope working at certain optical wavelength is limited by diffraction, which images a point source into a finite-sized spot. When two identical point sources are imaged and the spatial separation between their images is smaller than the spatial spread, the microscope cannot resolve them as two distinct spots. The minimum resolvable distance is therefore defined as the smallest separation between two point sources such that they can still be identified as two points. The intensity distribution of a point source is described by the Airy function as follows [94]:

$$I(r) = I_0 \left(2J_1 \left(\frac{2\pi}{\lambda} r \, NA \right) / \left(\frac{2\pi}{\lambda} r \, NA \right) \right)^2, \tag{3.29}$$

where I_0 is a constant, J_1 represents the first-order Bessel function, λ is the wavelength of light, r is the spatial radial coordinate, and as before NA is the short form for "numerical aperture," which is defined as the sine of half the angle at which rays are focused from the imaging lens when this is illuminated with a plane wave, multiplied by the refractive index of the object space. In Eq. 3.29, the peak intensity (at $r = 0$) is normalized and the first zero position of it is at $r = 1.22\lambda/NA$. According to Rayleigh's criterion, when the separation between two identical diffraction-limited spots is less than the distance from the center to the first zero position in the intensity distribution function, the two spots cannot be resolved. The minimum resolvable distance is therefore given by

$$\Delta x_R = 0.61\lambda/NA, \tag{3.30}$$

in which the intensity at the center of the overlapping region between two spots separated by Δx_R reaches about 80% of the center intensity of each spot. Sparrow's criterion [95], which is also commonly used, defines the smallest resolvable distance as the full width at half-maximum (FWHM) of the diffraction-limited peaks and is given by

$$\Delta x_S = 0.51\lambda/NA. \tag{3.31}$$

Experimentally, it is convenient to use Sparrow's criterion to determine the resolution of an optical system, since all that needs to be measured is the spot size (FWHM) in the focal plane from one point source. Note that this is only good for non-confocal imaging. For confocal imaging, the resolution criteria are different [95]. This criterion represents the best possible resolution for a staring optical system. Scanning systems can improve this resolution limit by an additional 20%–30% [96–98].

A diffraction-limited optical system is one in which all other limits to the resolution are smaller than the diffraction limits. Most modern objective lenses are diffraction-limited in the center of their field of view (FOV).

For optical probing, improvement in resolution can be made only by increasing the numerical aperture [99]. Since a sine function (in the definition of the NA) cannot exceed unity, and is already in the region of ~ 0.90 for the highest resolution objectives available, there is not much gain in increasing the marginal angle. The greatest benefit is gained by increasing the refractive index through which the marginal rays travel. This has been done through the use of liquid immersion objectives, and more recently by solid immersion lenses. Figure 3.8 shows experimentally obtained comparisons with and without the use of solid immersion lens (SIL) in the time resolved light emission

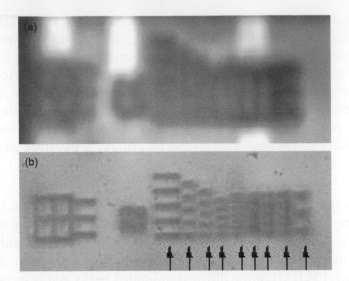

Figure 3.8 Image taken without (a) and with (b) solid immersion lens by the TRLEM system (100×). This is an array of diffusions with increasing distances between the lines. From the image, it appears that without the immersion lens one can observe two separated active lines only if they are separated by more than 1.3 um. With the immersion lens, it is possible to start to resolve the diodes when the distance is above 0.3 um.

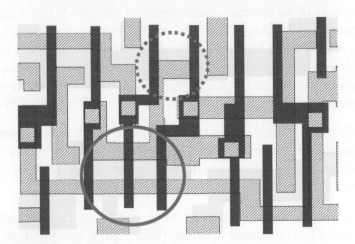

Figure 3.9 Schematic demonstration of the optical limitations inherent to IR analytical probing systems. The gray circles mark the laser scanning microscope (LSM) beam diameter (Full Width Half Maximum) without (solid circle) and with (dotted circle) immersion lens. In the CAD schematics taken from a 45 nm process, we can observe the logic indifference collection area measured using IC diagnostics tools.

microscope (TRLEM) system [97,98]. The improvement in resolution certainly makes the orientation easier; however, even with this improvement the Rayleigh criterion still does not allow resolution of the dense areas, as demonstrated in Fig. 3.9.

Another aspect that should be considered when increasing the numerical aperture is signal to noise ratio (the ratio between the photons of information coming from the object and the quantum noise they generate). Measurements done with the immersion lens using the laser voltage probe (LVP) technique [96] show that the signal—the collected photons related to the information (i.e., the reflection from the features of the 2-D structure)—was not degraded due to the incorporation of the lens. In the case of small features, the initial probing noise level is even better due to the localization of the beam.

A practical alternative to the conventional SIL is the diffractive lens. This lens is fabricated directly on the backside of the silicon of the device under test (DUT). This lens works on principles of diffractive optics and is around 250 nm thick. In commercial infrared (IR) microscope objective, the lens shows diffraction-limited resolution [100]. It should be noted that the lens works only for monochromatic light.

A perfect diffraction-limited optical system may still be unable to produce diffraction-limited images. The object being imaged is part of the optical system and can also degrade the total resolving power. For backside imaging, the curvature and the roughness of the back surface of the silicon can significantly reduce image quality. However, even a perfectly polished sample can introduce aberrations in the image quality at high numerical apertures. Aberrations (e.g., coma and astigmatism) will occur as the image spot moves away from the optical center, therefore mainly affecting the image quality and not the probing capability. Figure 3.9 demonstrates the optical limitation that is inherent to the IR analytical probing system with and without an immersion lens [101]. The gray circles mark the unresolved logic region without (solid circle) and with (dotted circle) immersion lens. In the CAD schematics taken from a 45 nm process, we can observe the logic indifference collection area measured using IC diagnostics tools. Any signal collected from this region can originate from one of the encircled transistors.

3.2.2 Super Resolving Algorithm
The algorithm suggested in this work is based on the Gerchberg–Papoulis algorithm. Its schematic sketch appears in Fig. 3.10.

The starting point of the basic algorithm assumes that we possess a high-resolution (HR) image (either a test object at high resolution or the layout plan of the electronic circuit) and a low-resolution (LR) image

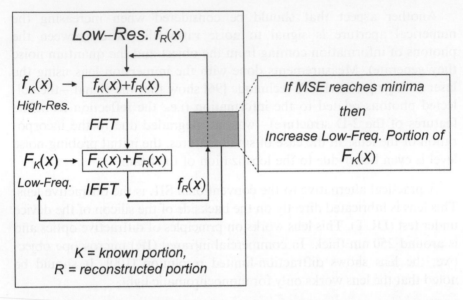

Figure 3.10 Schematic sketch of numerical superresolution algorithm. FFT and IFFT are the command names for the numerical realization of DFT and IDFT, respectively, in MathWorks © Matlac®. MSE stands for mean square error.

(either the test object at low resolution or the actual electronic circuit as captured by a low-resolution camera, respectively). Next the HR and LR images are aligned and resized to occupy the same sized matrix.

At first we divide each image into two vertical regions. Then, a new image is generated by combining the bottom region of the HR image and the top region of the LR image. Next, a Fourier transform is performed (actually a discrete Fourier transform, i.e., DFT) by implementing the FFT algorithm. The Fourier image obtained contains data from both halves of the new image. Since the lower frequencies are present in the LR image, we impose the lower frequencies from the Fourier transform of the original LR image (before it was spatially cut down by half). Next, an inverse Fourier transform is performed (actually an inverse DFT, i.e., the IDFT) by implementing the IFFT algorithm. At this stage, we replace the top region of the image by the top region of the HR image (in contrast to the previous case) and keep the bottom region. We again perform a Fourier transform to impose lower frequencies, and so on.

The basic algorithm comes to an end when the difference between images obtained in consecutive iterations is below a certain predefined threshold. At the final stage, the top region of the reconstructed image is taken from the iteration that began imposing the HR bottom region, and the bottom region is taken from the previous iteration (resulting from imposing the HR top part). The result obtained is actually the original LR image with emphasis at spatial locations where higher frequencies are expected. This process will become clearer on examination of Fig. 3.10.

The algorithm is upgraded by adding a dynamic parameter that is the amount of the lower frequencies imposed at the Fourier plane. Since imposing no data from the LR image leads to a nonconvergent procedure, and since imposing all of the LR frequencies at the initial stage might cause the procedure to converge to an undesirable local minimum, the algorithm imposes a small portion of the LR frequencies at the first stage. This portion remains the same until the correlation coefficient between the HR image and the reconstructed image increases no further. At this stage, the amount of the imposed LR frequencies only slightly increases. This upgraded algorithm results in an image that extracts the higher frequencies' data from the LR image using the HR companion image.

3.2.3 Experimental Results

We have applied the proposed numerical approach to several experimentally extracted images of resolution targets, as well as to real microelectronic chips.

In Fig. 3.11, one may see an example in which the algorithm was applied over a resolution target. The analyzed features (both horizontal and vertical) are 0.6 um in width and separated by 0.6 um. The laser LSM is 1064 nm. The device was thinned, polished to 100 um, and coated with anti-reflective coating. We used a 100 × objective with 0.85 NA. Figure 3.11(a) shows the layout of the target. In Fig. 3.11(b), we show the experimentally imaged object (lower part) captured by an infrared laser scanning microscope (LSM), and the result is obtained after using the digital superresolution algorithm (upper part). One may see that, in the superresolution image, the borders of the high resolution group appearing in the central lower part of Fig. 3.11(a) are very sharp and have high resolution, although the details of that group are too small to be seen. In the low resolution image, the borders of this group are blurred and nonvisible. The details of the resolution group appearing in the right central part of Fig. 3.11 (a) are resolved (both vertical and horizontal lines) in the superresolution image, while they are blurred in the low-resolution image. All of

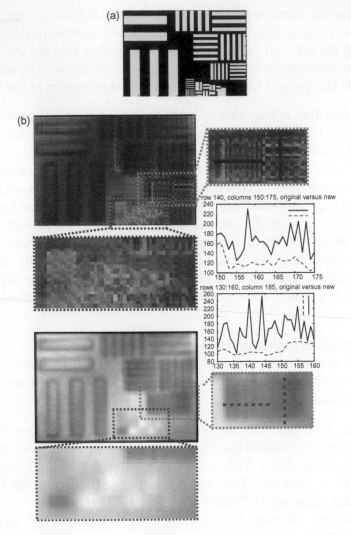

Figure 3.11 (a) High resolution reference layout image of a resolution target. (b) Experimentally imaged object (lower part) captured by infrared microscope (LSM) and the result obtained after digital superresolution algorithm (upper part). In the horizontal cross section, the contrast rises from 0.008 to 0.17. In the vertical cross section, the contrast rises from 0.019 to 0.26.

this is demonstrated numerically in the obtained enhanced contrast of the presented cross sections (vertical and horizontal). The contrast definition that we used for our computations is:

$$C = (I_{max} - I_{min})/I_{max} + I_{min} \qquad (3.32)$$

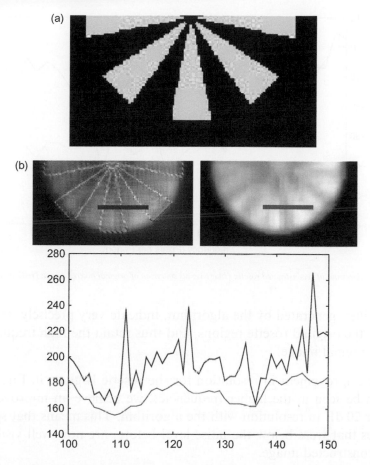

Figure 3.12 (a) High resolution reference layout image of resolution target. (b) Experimentally imaged object (right) captured by infrared microscope (SLM) and the result obtained after using the digital superresolution algorithm (left). The contrast in the cross section is increased from 0.093 to 0.175.

In Fig. 3.12, one may see another experimental example applied to a different resolution target (a rosette target). One may see that in the superresolution image, the borders of the rosette are high resolution, while they are completely blurred in the low-resolution image. All of this is demonstrated in the enhanced contrast of the presented cross sections (horizontal).

Although Figs. 3.11 and 3.12 indicate an improvement in contrast, it is important to note that this is actually an improvement in high frequency representation. This is clearest in Fig. 3.13, where the contrast peaks are not related to the original image, but rather to the transitions in that image (i.e., high frequencies). In the reconstructed figure, the

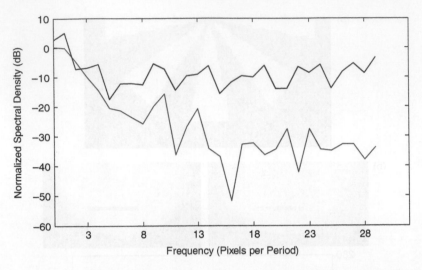

Figure 3.13 Spectrum of reconstructed rosette (blue) versus spectrum of original blurred image (red) in dB.

sharp lines, generated by the algorithm, indicate very precisely the borders of the original rosette regions, and thus retain the high frequencies of the original image.

The improvement in resolution for the rosette is given in Fig. 3.13. As can be seen in the higher frequencies, we achieve an improvement of over 20 dB in resolution with the algorithm. This means that spatial features that were below the noise level before, are now well visible in the reconstructed image.

In Fig. 3.14, we applied the proposed approach to experimentally imaged pictures of a real microelectronic chip. One may see that in the superresolution image, the CMOS elements' blocks appearing in the central lower part are much more separable than in the low resolution image. In the central upper part, one may see that in the superresolution image the four vertical blocks are visible and separable, while being completely blurred in the low resolution image.

The horizontal details of the square in the upper part of Fig. 3.14(b) are also completely non visible in the low resolution image, while visible in the superresolution image. All of this is demonstrated in the enhanced contrast of the presented cross sections (vertical and horizontal).

The last step in the demonstration of the applicability of the proposed approach, includes the comparison between the proposed approach and

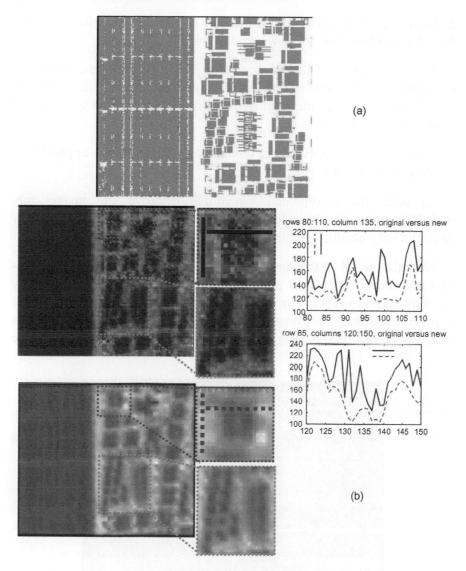

Figure 3.14 (a) High resolution reference layout image of random logic in a microelectronics chip. (b) Experimentally imaged object (lower part) captured by infrared microscope and the result obtained after applying digital superresolution algorithm (upper part). In the horizontal cross section, the contrast rises from 0.08 to 0.21. In the vertical cross section, the contrast is increased from 0.0359 to 0.117.

other commercially available algorithms. For comparison, the test image of Fig. 3.11 was reconstructed using three image enhancement algorithms present in commercial software. First, we used Corel® *PaintShop Pro* Ver. 9 (www.corel.com). Next, we used Imagetask® *Image: Fix and*

Enhance Ver. 1.52 (www.imagetask.com). Finally, we used Kodak®
Kodak EasyShare Ver. 6.4 (www.kodak.com).

The obtained results are given in Fig 3.15. As seen, the commercial
algorithms restore the mid-frequencies slightly better than the suggested algorithm, but the high-frequency data is restored only by the
algorithm suggested here.

In Fig. 3.16, we repeated the comparison, but this time applied it to
the rosette image of Fig. 3.12.

As seen, the commercial algorithms improve the image only
slightly, but the algorithm introduced here restores the high-frequency
data in a very accurate manner.

Note that in all of the simulations presented here, the image processing was done using offline processing via a MATLAB code written by
the authors. However, the number of product operations that are
required in order to converge to a reconstruction of an N by N pixels
image is about $2000 N^2 \log_2 N$. This number is obtained due to the

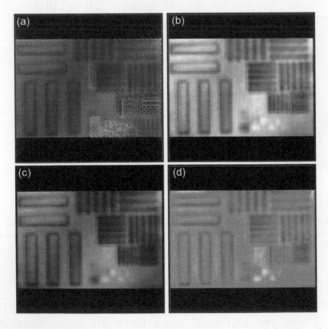

Figure 3.15 Comparing several image enhancement algorithms: (a) the suggested algorithm, (b) PaintShop Pro
algorithm for clarity and contrast enhancement, (c) Image: Fix and Enhance *algorithm for saturation and contrast enhancement, and (d)* Kodak EasyShare *algorithm for image and contrast enhancement.*

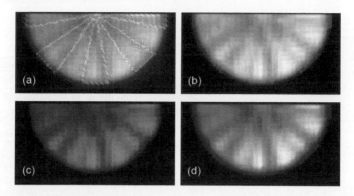

Figure 3.16 Comparing several image enhancement algorithms: (a) the suggested algorithm, (b) PaintShop Pro *algorithm, (c)* Image: Fix and Enhance *algorithm, and (d)* Kodak EasyShare *algorithm.*

fact that we have about 250 iterations while each iteration includes 2 FFT and 2 IFFT operations. Thus, for an image of N = 128 pixels, the number of product operations is approximately 200×10^6. Thus, the numerical requirements of the proposed algorithm are not very high, since even the processor of a regular cellular camera has 200 MIPS (million instructions per second), which means that the proposed processing can be performed in approximately 1 second. Nowadays, personal computers may reach tens of thousands of MIPS. Thus, if properly programmed, the algorithm will run for less than 10msec on such a processor. Therefore, although currently not in real time, this algorithm can be made a real-time one.

Since we attempt to address 45 nm technology of chips and beyond, we present in Fig. 3.17 the demonstration of the experimental acquisition of such an image. In the left upper part of the figure, we present the layout of the chip and in the left lower part we present the image that was acquired with a solid immersion lens from a 45 nm process, showing the resolution limit.

In the right part of the figure, we zoomed-in on a section of the chip and applied the proposed algorithm on it. The upper zoomed-in image is of the layout and the lower one is from the experimentally obtained image. The middle zoomed-in image is the one that we obtained with our algorithm. There is significant improvement in comparison to the image captured with the immersion lens. For instance, the squares of the layout appearing in the center of the reconstructed image are visible, while they are completely invisible in the experimentally captured image.

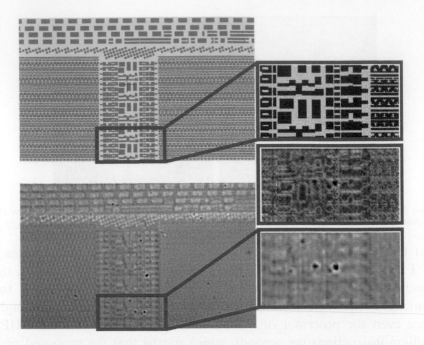

Figure 3.17 The layout and the experimentally acquired image with a solid immersion lens from a 45 nm process, showing the resolution limit. On the right, zoomed-in parts, we also present the image obtained using the proposed algorithm (in the middle).

3.3 USAGE OF RADON TRANSFORM FOR IMPROVED FAILURE ANALYSIS

3.3.1 The Radon Transform theory

In the Radon transform, multi-perspective 2-D snapshots of 3-D objects can be generated. The projection data that is gathered from multiple directions is used to reconstruct the 3-D detail of an object by using the Radon transform, which is a well-known tool. We discuss only the 2-D Radon transform here, since the 3-D version is a straight-forward generalization.

A special treatment is done in Ref. [102], addressing an example with a low number of projections (<300). We can avoid this by taking a sufficient number of samples. The Radon transform (RT) of a distribution $f(x, y)$ is given by:

$$P(r, \vartheta) = \int_{-\infty}^{\infty} \int f(x, y) \cdot \delta(x \cdot \cos(\theta) + y \cdot \sin \vartheta - r) dx dy \qquad (3.33)$$

The coordinate system for the RT is demonstrated in Fig. 3.18. The function $P(r, \vartheta)$ is also known as the sonogram. The resolution of the integral for each line will be determined by the system's resolution. Thus, the best focused image will affect not only the integral, but also the axial resolution along the entire integrated line.

We should add the resolution limit imposed by the imaging optics into the equation. Let us define the PSF of the optical imaging system as PSF(x,y):

$$G(r, \vartheta) = \int_{-\infty}^{\infty} (f(x, y) \otimes PSF(x, y) \cdot \delta(x \cdot \cos(\theta) + y \cdot \sin \vartheta - r) dx dy.$$

(3.34)

Note that the Dirac delta function is multiplying over the original function f(x,y), after it has been convolved with the PSF, so the sampling of the Radon transform over a line (r, ϑ) is a sampling of a smeared picture.

The task of topographic reconstruction is to find $f(x, y)$ given $G(r, \vartheta)$. Numbers of suitable algorithms have been proposed as solutions for the 3-D reconstruction problem from 2-D parallel projections. Most of these algorithms are based on 3-D filtered back projection (FBP), in which each 2-D projection is first convolved with an appropriate 2-D filter and then back projected into the 3-D image volume [103–106].

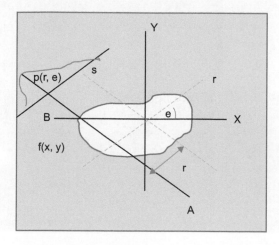

Figure 3.18 Coordinate system for the Radon transform.

Let us assume that we have a finite number of projections of an object which contains radioactive sources (Fig. 3.19). The projections of these sources at different degree intervals are represented on the sides of an octagon. This figure illustrates the basic idea behind back projection, which is simply to run the projections back through the image (hence the name "back projection") to obtain a rough approximation of the original object. The projections will interact constructively in regions that correspond to the emitting sources in the original image. A problem that is immediately apparent is the blurring (star-like artifacts) that occurs in other parts of the reconstructed image. One would expect that a high-pass filter could be used to eliminate blurring, and that is the case. The optimal way to eliminate these patterns in the noiseless case is through a ramp filter. The combination of back projection and ramp filtering is known as filtered back projection.

The formal formulation for the inverting Radon transform via filtered back projection operation is:

$$f(x, y) = \int\int_{-\infty}^{+\infty} F(\xi_x, \xi_y) \exp[i2\pi(\xi_x x + \xi_y y)] d\xi_x d\xi_y, \qquad (3.35)$$

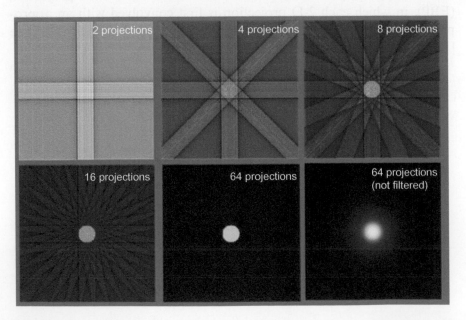

Figure 3.19 Filtered back projection.

Where F is the Fourier transform of the function f, and x, y are the lateral locations. While changing to polar coordinates (F_{polar}), we get:

$$f(x,y) = \int_0^{2\pi} \int_0^{\infty} F_{polar}(\xi, \theta)\exp[i2\pi\xi(x\cos\theta + y\sin\theta)]\xi d\xi d\theta, \qquad (3.36)$$

Where ξ is the distance from the center (radius), and θ is the angle from the horizontal axis.

$$f(x,y) = \int_0^{\pi} \int_{-\infty}^{\infty} |\xi|F_{polar}(\xi, \theta)\exp[i2\pi\xi(x\cos\theta + y\sin\theta)]d\xi d\theta, \qquad (3.37)$$

And replacing the term with the projection from Eq. (3.34) yields

$$f(x,y) = \int_0^{\pi} \left\{ \int_{-\infty}^{\infty} |\xi|G(\xi, \theta)\exp[i2\pi\xi(x\cos\theta + y\sin\theta)]d\xi \right\} d\theta, \qquad (3.38)$$

$$f(x,y) = \int_0^{\pi} \hat{g}(x\cos\theta + y\sin\theta, \theta)d\theta \text{ where } \hat{g}(s, \theta) = \int_{-\infty}^{\infty} |\xi|G(\xi, \theta)e^{i2\pi s\xi}d\xi.$$
$$(3.39)$$

Here, the filtering part in Eq. 3.39 is a forward Fourier transform for each of the angles. In the Fourier domain, the signal is high pass filtered with the filter $|\xi|$ while G is the Radon transform with the PSF impact as described in Eq. 3.34 and ξ is the radius on the polar coordinates, which provides the high pass filter (HPF) effect.

3.3.2 Failure analysis based upon Radon transform
Rapid advance of manufacturing processes continuously introduces reduced critical dimensions of features in modern integrated circuits. While the industry is highly motivated to introduce higher performance circuits, the investment in failure analysis tools is not comparable, especially during crisis periods. Normally, the development of new debugging tools lags behind analysis needs. For this reason, it is becoming essential to develop techniques that provide the capability to extend the applicability of older inspection tools to the shrinking dimensions of microelectronic devices.

For the vast majority of diagnostic tools, resolution features in the circuit remain the greatest challenge. The most significant leap in resolution improvement was achieved using solid immersion optics [99–101]. Nonetheless, common features in 45 nm processes are still 4 times smaller than the resolution limit. As of 2013, no significant development that can provide improved resolution over the silicon solid immersion capability, has been introduced.

Even a perfect diffraction limited optical system may still be unable to produce diffraction-limited images. The object being imaged is part of the optical system and can also degrade the total resolving power. For backside imaging, the curvature and the roughness of the back surface of the silicon can significantly reduce image quality. Therefore, any advantage optical systems might have does not reduce the need for some type of image recognition process.

The first limitation on accurate localization is driven by the difficulty of resolving small geometries from the backside. The active regions look like gray clouds. The use of accurate stages and computer aided design (CAD) overlay capability, which are incorporated into any modern backside tool, could alleviate spatial resolution issues. However, the current approach is to use overlay of the drawn polygons over the acquired image. This method requires some basic visual identification of layout features that can serve as anchors for the overlaid polygon image. The visual identification of the features is not trivial even for an expert analyst and any error could lead to incorrect analysis.

Integrated circuits are normally constructed out of straight and orthogonal lines. This means that certain projections of the image contain more data than others and lead us to propose the usage of the Radon transform, to be later explained. The main objective of this subsection is to present a new iterative image processing algorithm based on the Radon transform of optically collected images, in order to obtain a significant improvement in chip imaging resolution, and to then use this resolution improvement for failure analysis applications.

3.3.3 The Algorithm

In this subsection, we present a novel iterative algorithm that is based upon the Radon transform and which aims to improve the imaging resolution of microelectronic chips, leading to improved failure analysis.

The algorithm is based on the Gerchberg and Papoulis algorithm [92,93] and on the dynamic Gerchberg–Papoulis algorithm recently introduced [107]. Its schematic sketch appears in Fig. 3.20 [108].

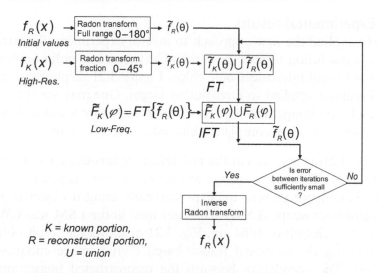

Figure 3.20 Schematic sketch of the numerical superresolution algorithm.

The starting point of the basic algorithm assumes that we possess the Radon transform of a high resolution (HR) image (the layout plan of the electronic circuit) at a small number of angles. We can also generate the Radon transform of a low resolution (LR) image we possess (the actual electronic circuit as captured by a low-resolution camera). At first a new image is generated by combining the Radon transform of the HR image at the given range of angles with the Radon of the LR image at all other angles.

Then a Fourier transform is performed. The obtained Fourier image contains data from both halves of the new image. Since the lower frequencies are present in the LR image, we impose the lower frequencies from the Fourier transform of the Radon transform of the original LR image. Next, an inverse Fourier transform is performed. At this stage, we replace the pixels referring to angles within the HR region with those of the HR image and keep the rest of the data. We again perform a Fourier transform to impose lower frequencies, and so on. The basic algorithm comes to an end when the difference between images obtained in consecutive iterations is below a certain predefined threshold. At the final stage, an inverse Radon transform is performed, and the reconstruction is completed.

3.3.4 Experimental results

We have applied the new approach to several experimentally extracted images of resolution targets (both with Cartesian and polar tendencies) as well as to real microelectronic chips. Figure 3.21 displays an experimental example applied to a resolution target. One may see that in the superresolution image, the borders of the rosette target are in high resolution, while they are completely blurred in the LR image.

In Fig. 3.21(a), we present the HR reference layout image of the resolution target. This rosette image was chosen due to its polar structure. In Fig. 3.21(b), we present the experimentally imaged object captured by infrared microscope (LSM). The laser used in the LSM was CW Nd: Yag at a wavelength of 1064 nm. Fig. 3.21(c) shows the result obtained after applying the proposed Radon-based digital image enhancement algorithm. The correlation between the reconstructed image and the

Figure 3.21 (a) High resolution reference layout image of a resolution target. (b) Experimentally imaged object captured by infrared microscope (LSM). (c) The result obtained after applying Radon-based super-resolving algorithm (93.6% correlation with (a)). (d) The result obtained after applying dynamic Gerchberg–Papoulis algorithm (53.1% correlation with (a)).

original HR image is 93.6%. For comparison, Fig. 3.21(d) shows the results obtained by applying the dynamic Gerchberg–Papoulis algorithm (given in Ref. [107], which always yields results equal to or better than the original Gerchberg–Papoulis algorithm); in this case the correlation coefficient is only 53.1%.

Since the authors aim to show that the Radon-based approach is valid for all linearly oriented images (whether polar or Cartesian), we performed the same test on a different resolution target, as shown in Fig. 3.22.

Again, Fig. 3.22(a) shows the HR reference layout image, this time a polar line target. Figure 3.22(b) shows the experimentally imaged object captured by infrared microscope (LSM). Figure 3.22(c) shows the result obtained after applying the Radon-based digital image enhancement

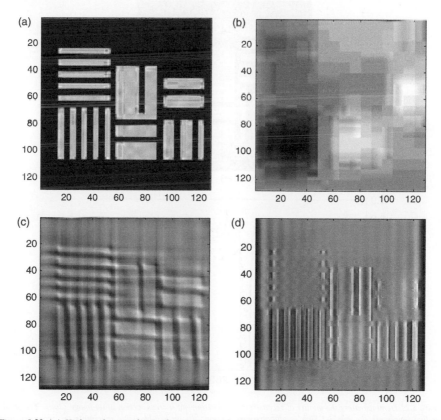

Figure 3.22 (a) High resolution reference layout image of a resolution target. (b) Experimentally imaged object captured by infrared microscope (LSM). (c) The result obtained after applying Radon-based superresolution algorithm (91.7% correlation with (a)). (d) The result obtained after applying the Gerchberg–Papoulis algorithm (41.3% correlation with (a)).

algorithm. The correlation between the reconstructed image and the original HR image is 91.7%. Figure 3.22(d) shows the results obtained by applying the original Gerchberg–Papoulis algorithm, and in this case the correlation coefficient is only 41.3%. When addressing contrast, the original blurred image has a contrast of only 0.254, while the new approach yields a contrast of 0.992.

Note the shearing effect in the Radon-based reconstruction is a result of using images with limited resolution and with an even number of pixels (thus the central pixel around which the Radon transformation is defined does not really exist).

To emphasize the quality of reconstruction using the new approach in comparison with the original Gerchberg–Papoulis algorithm, we present Fig. 3.23. The top part of Fig. 3.23 is the same image as that

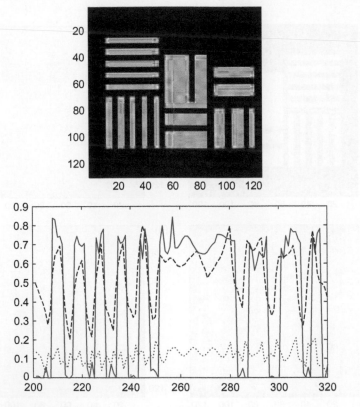

Figure 3.23 Demonstrating the quality of reconstruction (bottom) of a cross section of an image (top). The solid-red line represents the original high resolution cross section, the dashed-blue line represents the reconstruction using the Radon-based superresolution algorithm, and the magenta-dotted line represents the reconstruction using the Gerchberg–Papoulis algorithm.

shown in Fig. 3.22(a), but with the cross section test marked. The bottom part of Fig. 3.23 compares the desired image (solid-red) with the reconstruction using the Radon-based approach suggested here (dashed-blue) and the reconstruction using standard Gerchberg–Papoulis reconstruction (dotted-magenta).

It is clear from this plot that the Radon-based reconstruction follows the original image contour much better and has a much larger contrast than the standard previously published reconstruction technique.

It is important to notice two advantages of the new approach at this stage. First, as we are using the Fourier plane to impose restraints, a shifted version of the object will simply yield a shifted version of the reconstructed image. Second, since we are using a $0-180$ degree Radon transform, a rotated version of the object should yield the same results (while depending on the resolution of the Radon angles). These properties are demonstrated in Fig. 3.24. Figures 3.24(a) and 3.24(b) show a reconstructed image after shifting the original image by 10 pixels to the right, for the Radon-based approach and the original Gerchberg–Papoulis algorithm, respectively. While the correlation coefficient of the Radon-based approach changes by less than 0.11% (with respect to the one given in Fig. 3.22), the original Gerchberg–Papoulis algorithm results in a 13% change. Figures 3.24(c) and 3.24(d) show a reconstructed image after rotating the original image by 30 degrees clockwise, for the Radon-based approach and the original Gerchberg–Papoulis algorithm, respectively. While the correlation coefficient of the Radon-based approach changes by less than 0.12% (with respect to the example given in Fig. 3.22), the original Gerchberg–Papoulis algorithm results in a change of 16%.

Since this subsection attempts to address 45 nm technology of chips and beyond, we present in Figs. 3.25–3.27 the demonstration of the experimental acquisition of such images..

In Fig. 3.25(a), we present the layout of a certain chip and in the upper left part of Fig. 3.25(b), we present the experimentally acquired image obtained with a solid immersion lens (SIL) from the 45 nm process, showing the resolution limit. In Fig. 3.25(c), we show the improvement obtained after applying the proposed image-processing algorithm, and in Fig. 3.25(d) we show the improvement obtained after applying the dynamic Gerchberg–Papoulis algorithm.

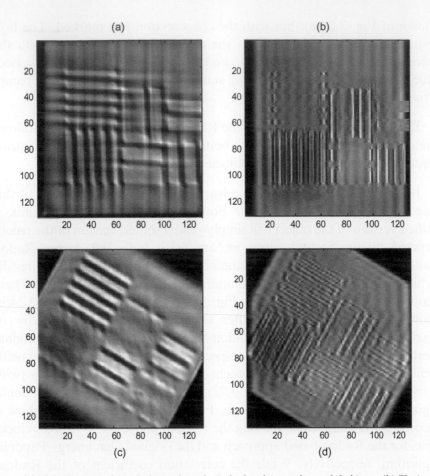

Figure 3.24 (a) The image obtained after applying the Radon-based approach to a shifted input. (b) The image obtained after applying the Gerchberg–Papoulis algorithm to a shifted input. (c) The image obtained after applying the Radon-based approach to a rotated input. (d) The image obtained after applying the Gerchberg–Papoulis algorithm to a rotated input.

The correlation coefficients between the obtained results in Figs. 3.25 (c) and 3.25(d) and the original HR image are 99.8% and 95.8% respectively, and one may notice that the left part of the image, which was completely lost in the original restoration algorithm, appears as quite a sharp image when using the new approach.

In Figures 3.26 and 3.27, we repeat the same demonstration while the image processing algorithm is applied over different portions of an experimentally captured image of a chip without poly (only active diffusion) layer. The marked rectangles demonstrate the reconstruction of different modules (in Fig. 3.26) and the obtained correlation coefficient

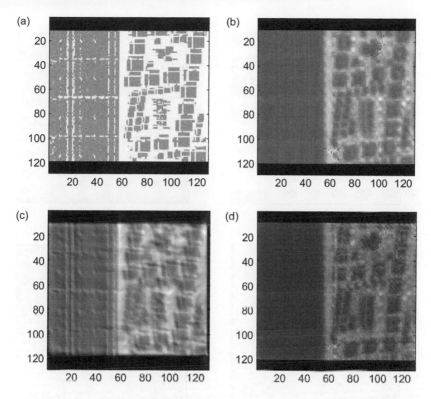

Figure 3.25 (a) The layout image of a 45 nm process chip. (b)The experimentally acquired image with solid immersion lens, showing the resolution limit. (c) The image obtained after applying the Radon-based image processing approach (99.8% correlation with (a)). (d) The result obtained after applying the dynamic Gerchberg–Papoulis algorithm (95.8% correlation with (a)).

is 99.8% for the suggested approach (while only 92.8% for the dynamic Gerchberg–Papoulis approach).

Figure 3.27 does not only show an improvement in detection, but also demonstrates in Fig. 3.27(d) a false detection obtained using the dynamic Gerchberg–Papoulis algorithm, which does not appear in the reconstruction suggested here (see the regions marked by the red rectangles). The correlation coefficients between the obtained results in Figures 3.27(c) and 3.27(d) and the original HR image are 99.8% and 92.6% respectively.

The major goal of the proposed technique is to overcome the resolution limits in backside IR imaging. Due to the poor resolution, the analyst is not able to correlate between the acquired images with the layout database. The proposed method can help to overcome this

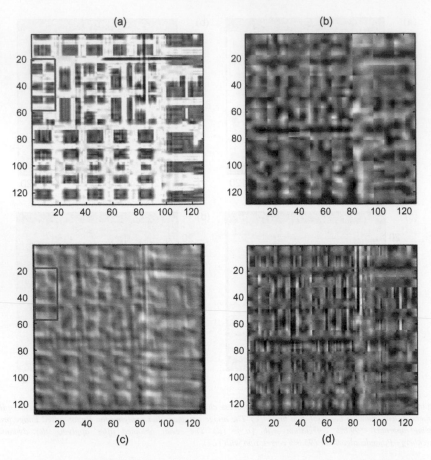

Figure 3.26 Layout with only active diffusion. (a) The layout image of a 45 nm process chip. (b) The experimentally acquired image with solid immersion lens, showing the resolution limit. (c) The result obtained after applying Radon-based superresolution algorithm (99.8% correlation with (a)). (d) The result obtained after applying the dynamic Gerchberg–Papoulis algorithm (92.8% correlation with (a)).

major limitation. The correlation is based on matching the phase coherence of the layout (expected) and the image (acquired), and thus any defect in the image will not match expected detail and hence will be treated as noise. When the analyst is looking for a specific defect, he or she can process the device using high-resolution tools (SEM or TEM) and examine the image for unexpected damage—in which case the IR resolution barrier is removed and the proposed technique is no longer needed. Nevertheless, Fig. 3.28 demonstrates how the algorithm keeps the defects in an image while improving the contour.

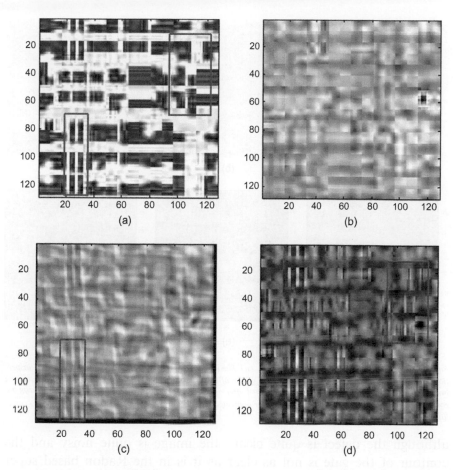

Figure 3.27 Layout with only active diffusion. (a) The layout image of a 45 nm process chip. (b) The experimentally acquired image with solid immersion lens, showing the resolution limit. (c) The result obtained after applying the Radon-based superresolution algorithm (99.8% correlation with (a)). (d) The result obtained after applying the dynamic Gerchberg−Papoulis algorithm (92.6% correlation with (a)).

In Fig. 3.28(a), one sees a layout of a chip with active fusion and poly. Figure 3.28(b) shows the experimentally acquired image of the chip using SEM. Figure 3.28(c) focuses on a specific gate in the chip, containing a defect appearing as an electrical-over stress (EOS), while Fig. 3.28(d) shows the same gate as captured in low resolution. Figure 3.28(e) indicates the result of the algorithm suggested here, leading to a 98.7% correlation coefficient with the original layout image and at the same time showing the defect. Figure 3.28(f) indicates the result of the dynamic Gerchberg−Papoulis algorithm, leading to a 94.2% correlation coefficient with the original layout image, and—

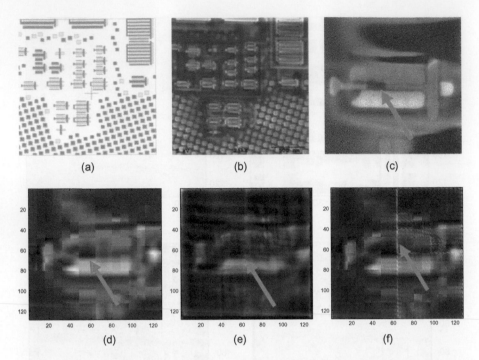

(a)   (b)   (c)

(d)   (e)   (f)

Figure 3.28 Layout with active diffusion and poly. (a) The layout image of a 45 nm process chip. (b) SEM inspection of the chip. (c) Zooming in: SEM inspection of one of the gates (arrow indicating defect). (d) Low resolution image of the inspected gate. (e) The result obtained after applying Radon-based superresolution algorithm (98.7% correlation with (c)). (f) The result obtained after applying dynamic Gerchberg—Papoulis algorithm (94.2% correlation with (c)).

although the defect is quite clear—the image is quite noisy and the contour of the gate is not as clear as it is in the Radon based super resolving algorithm.

REFERENCES

[1] Goodman JW. Introduction to Fourier optics. 2nd ed. New York: McGraw-Hill; 1996.

[2] Abbe E. Beitrage zur theorie des mikroskops und der mikroskopischen wahrnehmung. Arch Mikrosk Anat 1873;9:413–68.

[3] Zalevsky Z, Mendlovic D, Lohmann AW. Optical system with improved resolving power. Prog Opt 1999;XL: Ch. 4

[4] Zalevsky Z, Mendlovic D. Optical super resolution. Springer; 2002.

[5] Toraldo di Francia G. Supergain antennas and optical resolving power. Nuovo Cimento 1952;9(Suppl.):426–35.

[6] Fellgett PB, Linfoot EH. On the assessment of optical images. Phil Trans R Soc London Ser A 1955;247:369–407.

[7] Toraldo di Francia G. Resolving power and information. J Opt Soc Am 1955;45:497–501.

[8] Kartashev AI. Optical system with enhanced resolving power. Opt Spectra 1960;9:204–6.

[9] Lohmann AW, Paris DP. Superresolution for nonbirefringent objects. Appl Opt 1964;3:1037—43.

[10] Lukosz W. Optical systems with resolving powers exceeding the classical limits. J Opt Soc Am 1966;56:1463—72.

[11] Lukosz W. Optical systems with resolving powers exceeding the classical limits II. J Opt Soc Am 1967;57:932—41.

[12] Grimm MA, Lohmann AW. Superresolution image for one-dimensional object. J Opt Soc Am 1966;56:1151—6.

[13] Bachl A, Lukosz W. Experiments on superresolution imaging of a reduced object field. J Opt Soc Am 1967;57:163—9.

[14] Toraldo di Francia G. Degrees of freedom of an image. J Opt Soc Am 1969;59:799—804.

[15] Cox IJ, Sheppard JR. Information capacity and resolution in an optical system. J Opt Soc Am A 1986;3:1152—8.

[16] Den Dekker AJ, Van den Bos A. Resolution: a survey. J Opt Soc Am A 1997;14:547—57.

[17] Sales TRM, Morris GM. Fundamental limits of optical superresolution. Opt Lett 1997;22:582—4.

[18] Sheppard CJR, Larkin KG. Information capacity and resolution in three-dimensional imaging. Optik 2003;113(12):548—50.

[19] Sheppard CJR. Fundamentals of superresolution. Micron 2007;38:165—9.

[20] Lohmann AW, Dorsch RG, Mendlovic D, Zalevsky Z, Ferreira C. About space-bandwidth product of optical signals and systems. J Opt Soc Am A 1996;13:470—3.

[21] Mendlovic D, Lohman AW. SW-Adaptation and its application for super resolution: fundamentals. J Opt Soc Am A 1997;14:558—62.

[22] Mendlovic D, Lohman AW, Zalevsky Z. SW-Adaptation and its application for super resolution: examples. J Opt Soc Am A 1997;14:562—7.

[23] Wolf KB, Mendlovic D, Zalevsky Z. Generalized Wigner function for the analysis of superresolution systems. Appl Opt 1998;37:4374—9.

[24] Zalevsky Z, Mendlovic D, Lohmann AW. Understanding superresolution in Wigner space. J Opt Soc Am A 2000;17:2422—30.

[25] Francon M. Amélioration the résolution d'optique. Il Nuovo Cimento 1952;9 (Suppl.):283—90.

[26] Shemer A, Mendlovic D, Zalevsky Z, García J, García-Martínez P. Superresolving optical system with time multiplexing and computer decoding. Appl Opt 1999;38:7245—51.

[27] Armitage JD, Lohmann AW, Parish DP. Superresolution image forming systems for objects with restricted lambda dependence. Jpn J Appl Phys 1965;4:273—5.

[28] Bartelt H, Lohmann AW. Optical processing of 1-D signals. Opt Commun 1982;42:87—91.

[29] Gartner W, Lohmann AW. An experiment going beyond Abbe's limit of diffraction. Z Phys 1963;174:18—23.

[30] Zlotnik A, Zalevsky Z, Marom E. Superresolution with nonorthogonal polarization coding. Appl Opt 2005;44:3705—15.

[31] Zalevsky Z, García-Martínez P, García J. Superresolution using gray level coding. Opt Express 2006;14:5178—82.

[32] Irani M, Peleg S. Super resolution from image sequences. ICPR 1990;2:115—20.

[33] Park SC, Park MK, Kang MG. Super-resolution image reconstruction: a technical overview. IEEE Signal Proc Mag 2003;20(3):21−36.

[34] Zalevsky Z, Mendlovic D, Marom E. Special sensor masking for exceeding system geometrical resolving power. Opt Eng 2000;39:1936−42.

[35] Porter AB. On the diffraction theory of microscope vision. Philos Mag 1906;6(11):154−6.

[36] Nyquist H. Certain topics in telegraph transmission theory. Trans AIEE 1928;47:617−44. Reprint as classic paper in: Proc IEEE 2002;90(2):280−305.

[37] Shannon CE. Communication in the presence of noise, Proc Inst Radio Eng 1949;37:10−21. Reprint as classic paper in: Proc IEEE 1998;86(2):447−457.

[38] Fortin J, Chevrette P, Plante, R. Evaluation of the microscanning process, SPIE Vol. 2269, Infrared Technology XX, Andresen BF, editor. 1994; 271−279.

[39] Ashok A, Neifeld MA. Pseudo-random phase masks for super-resolution imaging from subpixel shifting. Appl Opt 2007;46:2256−68.

[40] Ben-Ezra M, Zomet A, Nayar SK. Video super-resolution using controlled subpixel detector shifts. IEEE Trans Pattern Anal Mach Intell 2005;27:977−87.

[41] Borman S, Stevenson R. Super-resolution from image sequences−A review. In: Proceedings of the 1998 Midwest Symposium on Circuits And Systems, Notre Dame, IN, USA; 1998: pp. 374−78.

[42] Bascle B, Blake A, Zisserman A. Motion deblurring and super-resolution from an image sequence. Proc Eur Conf Comput Vis 1996;2:573−82.

[43] Ben-Ezra M, Nayar SK. Motion-based motion Deblurring. IEEE Trans Pattern Anal Mach Intell 2004;26:689−98.

[44] Elad M, Feuer A. Restoration of single super-resolution image from several blurred, noisy and down-sampled measured images. IEEE Trans Image Proc 1997;6:1646−58.

[45] Elad M, Feuer A. Super-resolution reconstruction of continuous image sequence. IEEE Trans Pattern Anal Mach Intelligence (PAMI) 1999;21:817−34.

[46] Elad M, Hel-Or Y. A fast super-resolution reconstruction algorithm for pure translational motion and common space invariant blur. IEEE Trans Image Process 2001;10:1187−93.

[47] Zalevsky Z, Shamir N, Mendlovic D. Geometrical super-resolution in infra-red sensor: Experimental verification. Opt Eng 2004;43:1401−6.

[48] Fixler D, Garcia J, Zalevsky Z, Weiss A, Deutsch M. Pattern projection for subpixel resolved imaging in microscopy. Micron 2007;38:115−20.

[49] Borkowski A, Zalevsky Z, Javidi B. Geometrical super resolved imaging using non periodic spatial masking. JOSA A 2009;26:589−601.

[50] Park SC, Park MK, Kang MG. Super-resolution image reconstruction: a technical overview. IEEE Signal Proc Mag 2003;21−36.

[51] Katsaggelos AK, editor. Digital image restoration, vol. 23. Heidelberg, Germany: Springer-Verlag. Springer; 1991.

[52] Schoenberg IJ. Cardinal interpolation and spline functions. J Approx Theory 1969;2:167−206.

[53] Crochiere RE, Rabiner LR. Interpolation and decimation of digital signals—A tutorial review. Proc IEEE 1981;69(3):300−31.

[54] Unser M, Aldroubi A, Eden M. Enlargement or reduction of digital images with minimum loss of information. IEEE Trans Image Process 1995;4(3):247−58.

[55] Dvorchenko VN. Bounds on (deterministic) correlation functions with applications to registration. IEEE Trans Pattern Anal Mach Intell 1983;5(2):206–13.

[56] Tian Q, Huhns MN. Algorithm for subpixel registration. Computer Vision, Graphics, Image Proc 1986;35:220–33.

[57] Bernstein CA, Kanal LN, Lavin D, Olson EC. A geometric approach to subpixel registration accuracy. Comput Vis Graph Image Process 1987;40:334–60.

[58] Brown LG. A survey of image registration techniques. ACM Comput Surveys 1992;24 (4):325–76.

[59] Ur H, Gross D. Improved resolution from sub-pixel shifted pictures. CVGIP: Graph Model Im Proc 1992;54:181–6.

[60] Papoulis A. Generalized sampling theorem. IEEE Trans Circuits Syst 1977;24:652–4.

[61] Brown JL. Multi-channel sampling of low pass signals. IEEE Trans Circuits Syst 1981;CAS-28(2):101–6.

[62] Komatsu T, Aizawa K, Igarashi T, Saito T. Signal-processing based method for acquiring very high resolution image with multiple cameras and its theoretical analysis. Proc Inst Elec Eng 1993;140(1):19–25, Pt. I.

[63] Landweber L. An iteration formula for Fredholm integral equations of the first kind. Am J Math 1951;73:615–24.

[64] Komatsu T, Igarashi T, Aizawa K, Saito T. Very high resolution imaging scheme with multiple different-aperture cameras. Signal Processing: Image Commun 5; 1993, pp. 511–526

[65] Alam MS, Bognar JG, Hardie RC, Yasuda BJ. Infrared image registration and high-resolution reconstruction using multiple translationally shifted aliased video frames. IEEE Trans Instrum Meas 2000;49:915–23.

[66] Tsai RY, Huang TS. Multipleframe image restoration and registration. Advances in computer vision and image processing. Greenwich, CT: JAI Press Inc.; 1984, pp. 317–39.

[67] Bose NK, Kim HC, Valenzuela HM. Recursive implementation of total least squares algorithm for image reconstruction from noisy, undersampled multiframes. In: Proceedings of the IEEE Conference Acoustics, Speech and Signal Processing, Minneapolis, MN, 1993, vol. 5, pp. 269–72.

[68] Rhee SH, Kang MG. Discrete cosine transform based regularized high-resolution image reconstruction algorithm. Opt Eng 1999;38(8):1348–56.

[69] Hong MC, Kang MG, Katsaggelos AK. A regularized multichannel restoration approach for globally optimal high resolution video sequence. SPIE VCIP, vol. 3024. San Jose CA; 1997, pp. 1306–317.

[70] Hong MC, Kang MG, Katsaggelos AK. An iterative weighted regularized algorithm for improving the resolution of video sequences. Proc Int Conf Image Process 1997;2:474–7.

[71] Kang MG. Generalized multichannel image deconvolution approach and its applications. Opt Eng 1998;37(11):2953–64.

[72] Hardie RC, Barnard KJ, Bognar JG, Armstrong EE, Watson EA. High-resolution image reconstruction from a sequence of rotated and translated frames and its application to an infrared imaging system. Opt Eng 1998;37(1):247–60.

[73] Tom BC, Katsaggelos AK. Reconstruction of a high-resolution image by simultaneous registration, restoration, and interpolation of low-resolution images. In: Proceedings of the 1995 IEEE International Conference on Image Processing, vol. 2. Washington, DC; 1995, pp. 539–542.

[74] Schultz RR, Stevenson RL. Extraction of high-resolution frames from video sequences. IEEE Trans Image Process 1996;5:996–1011.

[75] Stark H, Oskoui P. High resolution image recovery from image-plane arrays, using convex projections. J Opt Soc Am A 1989;6:1715–26.

[76] Irani M, Peleg S. Improving resolution by image registration. CVGIP: Graph Model Im Proc 1991;53:231–9.

[77] Elad M, Feuer A. Superresolution restoration of an image sequence: adaptive filtering approach. IEEE Trans Image Process 1999;8:387–95.

[78] Rajan D, Chaudhuri S. Generation of super-resolution images from blurred observations using an MRF model. J Math Imag Vis 2002;16:5–15.

[79] Rajan D, Chaudhuri S. Generalized interpolation and its applications in super-resolution imaging. Image Vis Comput 2001;19:957–69.

[80] Joshi MV, Chaudhuri, S. Super-resolution imaging: Use of zoom as a cue. In: Proceedings of the Indian Conference Vision, Graphics and Image Processing, Ahmedabad, India; 2002, pp. 439–44.

[81] Wirawan W, Duhamel P, Maitre H. Multi-channel high resolution blind image restoration. In: Proceedings of IEEE ICASSP, AZ, Nov. 1989, pp. 3229–232.

[82] Hadamard J. Lectures on Cauchy's problem in linear partial differential equation. New York: Dover; 1923.

[83] Tekalp AM, Ozkan MK, Sezan MI. High-resolution image reconstruction from lower-resolution image sequences and space varying image restoration. Proceedings of IEEE International Conference Acoustics, Speech and Signal Processing (ICASSP) 1992;3: 169–72.

[84] Kim SP, Bose NK, Valenzuela HM. Recursive reconstruction of high resolution image from noisy undersampled multiframes. IEEE Trans Acoust 1990;38(6):1013–27.

[85] Bose NK, Kim HC, Valenzuela HM. Recursive total least squares algorithm for image reconstruction from noisy, undersampled frames. Multidim Syst Sign Proc 1993;4 (3):253–68.

[86] Kim SP, Su WY. Recursive high-resolution reconstruction of blurred multiframe images. IEEE Trans Image Process 1993;2(4):534–9.

[87] Tikhonov A, Arsenin VY. Solutions of ill-posed problems. Washington, DC: V.H. Winston and Sons; 1977.

[88] Hong M, Kang MG, Katsaggelos AK. An iterative weighted regularized algorithm for improving the resolution of video sequences. IEEE Int Conf Image Process 1997;2:474–7.

[89] Ozaktas HM, Zalevsky Z, Kutay MA. The fractional Fourier transform with applications in optics and signal processing. John Wiley and Sons; 2001.

[90] Zalevsky Z. Wigner based analysis of geometric related resolution degradation and geometric super resolution configuration. In: Proceedings of the progress in electro-magnetic research symposium (PIERS), PIERS Online, 2011;7(5): 451–55.

[91] Gerchberg RW, Saxton WO. A practical algorithm for determination of phase from image and diffraction plane picture. Optik (Stuttgart) 1972;35:237–46.

[92] Gerchberg RW. Super-resolution through error energy reduction. Opt Acta (Lond) 1974;21 (9):709–20.

[93] Papoulis A. A new algorithm in spectral analysis and band-limited extrapolation. IEEE Trans Circuits Syst 1975;22(9):735–42.

[94] Born M, Wolf E. Principles of optics. New York: Pergamon; 1980, pp. 253, 468

[95] Corle TR, Kino GS. Confocal scanning microscopy and related imaging systems. New York: Academic Press; 1996.

[96] Paniccia M, Eiles T, Rao VRM, Wai Mun Y. Novel optical probing technique for flip chip packaged microprocessors. Test Conference International Proceedings, 18−23 Oct 1998.

[97] Vickers J, Pakdaman N, Kasapi S. Prospects of time-resolved photon emission as a debug tool. Proceedings of the 28th ISTFA; 2002.

[98] Tsang JC, Kash JA. Temporal characterization of CMOS circuits by time resolved emission microscopy. 55th Device Research Conference Digest; 23−25 Jun 1997, pp. 24−25.

[99] Falk RA. Optimizing backside image quality. Proceedings of the 28th ISTFA; 2002.

[100] Zachariasse F, Goossens M. Diffractive lenses for high resolution laser based failure analysis. Proceedings of the 28th IPFA; 2006.

[101] Weizman Y, Baruch E. VLSI design for functional failure analysis in the <90 nm and flip-chip era. Proceedings of the 29th ISTFA; 2003.

[102] Simonov E. Use of image reconstruction algorithms based on the integral radon transform in small angle X ray computer tomography. Biomed Eng 2004;38(6):287−91.

[103] Chesler A, Pelc NJ. Utilization of cross-plane rays for 3D reconstruction by filtered back-projection. J Comput Assist Tomogr 1979;3:385−95.

[104] Agi I, Hurst PJ, Current KW. VLSI signal processing IV. A pipelined architecture for radon transform computation in a multiprocessor array. New York: IEEE Press; 1990.

[105] Bracewell RN. Two-dimensional imaging. Englewood Cliffs, NJ: Prentice Hall; 1995, pp. 505−37.

[106] Lim JS. Two-dimensional signal and image processing. Englewood Cliffs, NJ: Prentice Hall; 1990, pp. 42−5.

[107] Gur E, Weizman Y, Zalevsky Z. Superresolved imaging of microelectronic devices for improved failure analysis. IEEE Trans Device Mater Reliab 2009;9(2).209−14.

[108] Gur E, Weizman Y, Perdu P, Zalevsky Z. Radon transform based image enhancement for microelectronic chips inspection. IEEE Trans Device Mater Reliab 2010;10:403−8.

Printed and bound by CPI Group (UK) Ltd, Croydon, CR0 4YY

03/10/2024

01040420-0014